For a Science
of Social Man

THE MACMILLAN COMPANY
NEW YORK · CHICAGO
DALLAS · ATLANTA · SAN FRANCISCO
LONDON · MANILA

BRETT-MACMILLAN LTD.
TORONTO

edited *by* JOHN GILLIN

For a Science
of Social Man

CONVERGENCES IN ANTHROPOLOGY, PSYCHOLOGY, AND SOCIOLOGY

by HOWARD BECKER
JOHN GILLIN
A. IRVING HALLOWELL
GEORGE PETER MURDOCK
THEODORE M. NEWCOMB
TALCOTT PARSONS
M. BREWSTER SMITH

THE MACMILLAN COMPANY   NEW YORK

Preface

THIS VOLUME WAS made possible, in the first instance, by a modest financial grant to John Gillin in 1949 from the Wenner-Gren Foundation for Anthropological Research (then called the Viking Fund, Inc.). The original purpose of the grant was "to explore the possibilities of interdisciplinary integration in the human or man sciences." At that time it was thought that a week-long conference of a few leading experts might produce tangible progress toward such a goal. During the academic year 1949–1950 an interdisciplinary faculty seminar was conducted at the University of North Carolina, partly for the purpose of testing out the conference idea, and devoted to a consideration of concepts common to several of the disciplines. Aid in the form of research assistants was supplied by the Institute for Research in Social Science of the University of North Carolina. Among the senior participants were Nicholas J. Demerath, Louis O. Kattsoff, Harold G. McCurdy, George E. Nicholson, Jr., Rupert B. Vance, and Karl Zener (the latter from Duke University). A large amount of theoretical material, plus recorded discussions, was accumulated and filed, but at the end of the academic year most of the members were scattered about this country and abroad by responsibilities for research projects and other obligations, so that the seminar was perforce discontinued. Also, as a result of experience in the seminar it became increasingly doubtful that a short conference of experts or authorities, however eminent, would at this point be able to produce anything constructive toward the goal in mind. Accordingly, after a period of reflection and consultation, the present plan was decided upon. The authors have held two conferences in

New York, the first to agree upon a plan for the book, and the second to go over together the prepared chapters. Each author has, however, prepared his own chapter at his leisure (if that is the word) rather than in the heat of discussion or debate.

Profound thanks are recorded to the Wenner-Gren Foundation for Anthropological Research and to its Director of Research, Dr. Paul Fejos, for their enlightened policy which made the preparation of this book possible. The Foundation has not, however, provided the costs of publication, and the Foundation of course is not necessarily in agreement with any of the statements made herein. The editor also wishes to state his indebtedness to the Institute for Research in Social Science of the University of North Carolina and to Dr. Gordon W. Blackwell, its director, for active interest and material aid in carrying out this project.

The manuscript was closed on June 15, 1953.

JOHN GILLIN

Chapel Hill, N.C.

Contents

For a Science
of Social Man

CHAPTER I

Grounds for a Science of Social Man

JOHN GILLIN

ALMOST EVERYONE WHO has given thought to the matter seems to agree that increased coordination and sharper focusing of our knowledge of man's behavior in society is desirable and necessary. If science has been able to discover and organize the means whereby human beings can now exterminate the human species, the argument runs, why should it not also be able to marshall its resources to insure survival? And why not, indeed? Naïve as it may sound to some, this is quite a legitimate question.

Nevertheless opinions on this matter vary. On the one hand, we meet those who solemnly assure us that, desirable as it might be, a coordinated science capable of handling the larger human issues of the modern world is a dream, a product of wishful thinking. Human nature and human relations, they tell us, are inherently of such a sort that they can never be handled by science. To many, including the present authors, such advice is a counsel of despair that, when not based on ulterior or extraneous motives, is derived from inadequate knowledge of either man or science.

The opposite camp insists that a science of social man *is* possible and that no effort should be spared in perfecting it. But some who are of this opinion are inclined to think that any measurable progress lies a long way in the future, because, they claim, the disciplines dealing with human behavior and its products are still "too young and immature."

The authors of this book, in common with many other specialists and scientific organizations, have faith in the application of science to human affairs, and they do not feel that either the alleged impossibility of a science of human social behavior or the supposedly inevitable delay in its development is to be taken for granted without a close inspection of the situation and a careful consideration of alternatives and possibilities. We believe that a better communication of extant theory and knowledge across disciplinary lines will not only pave the way for a more comprehensive approach to human problems, but will also stimulate the scope and predictive power of behavior science, including all the specialties involved. Yet we have felt that something more constructive is needed than merely a collective blast of exhortation. There is little to be gained by adding our voices to the lamentations and scoldings of those who angrily demand that social science immediately "catch up" with nuclear physics.

If the sciences that are supposed to illuminate the behavior of man, his societies, and his cultures are to be collectively rather than separately helpful to mankind in its uneasy state, we must first of all, it would seem, understand what these disciplines have in common.

As a modest step in this direction we have thought it most convenient to begin with an examination of certain of the mutual problems of those three "human science disciplines" that have come to be considered the "core" sciences of human behavior in society—namely, anthropology, psychology, and sociology. Such an enterprise is rendered the more expedient by the fact that during the past twenty-five years or so the followers of these specialties have, sometimes in spite of everything and often a trifle awkwardly, found themselves increasingly in each other's company, rather like a collection of Protestants, Jews, and Catholics who keep turning up at the same cocktail parties and who are, at first, put to it to make polite conversation, and then gradually drawn into discussing fundamental issues. The anthropologists, psychologists, and sociologists have repeatedly discovered each other in the same situation and have had to come to some sort of *modus vivendi*. And out of this has emerged the realization that they do have many mutual problems and much common knowledge, even though approached from different points of view, and that joint work might for certain purposes prove more rewarding than the independent role.

Accordingly we have undertaken on the following pages to bring to light certain agreements and convergences, especially in theory, among these sister disciplines and to point to promising possibilities that will, in our opinion, contribute to the further development of a science of social man.

At the same time we do not advocate, in fact we reject, any form of scientific authoritarianism such as is implied in some schemes for the "unification of the sciences." In approaching the possibilities of interdisciplinary collaboration, we propose, if the figure be appropriate, not a Monolithic State, but rather a Federal Union of the specialties dealing scientifically with human behavior in society. In such a Federal Union the several member disciplines would be able to pool their scientific resources for the solution of certain problems requiring multidisciplinary treatment, while maintaining a species of "states' rights" that would guarantee full freedom for each member to attend to concerns that seem to be of more specialized interest. We feel that useful convergence in theory at certain points and practical collaboration in some types of research and its application need in no way impinge upon the autonomy of the established disciplines nor block that necessary search for new truth which can only be achieved by those equipped with theoretical guidance and technical knowledge peculiar to their respective specialties. Indeed it is entirely possible that, as the problems of a cooperative science of social man become clearer, it will be seen that more rather than fewer specialisms are needed. At any rate, it is primarily to the elucidation of some of the theoretical bases upon which a Federal Union of the sciences of social man could be erected that the present book is devoted.

We must, nonetheless, admit that certain critics are not entirely wrong when they say that one of the intellectual curiosities of mid-twentieth century United States (and perhaps of Western civilization generally) is an impression many have. This is that, whereas our knowledge of "nature" and of material processes has seemingly become increasingly consolidated and integrated, our understanding of man in society has tended toward fractionation into semi-autonomous (and sometimes mutually jealous) disciplines with a consequent obfuscation of thought and theory that passeth all understanding. Something of this point of view prevailed, for example, when the United States Congress in establishing the National Science Foundation in 1950, refused to include the "so-called social sciences" explicitly in the plan (and, of course, in the appropriations), partly in the belief that they really are not sciences at all.

And the late John Dewey, often called America's most "social-minded" philosopher, made the following gloomy pronouncement in one of his last published contributions.

. . . Inquiries in the human field, in the so-called "social" sciences, are now in as backward a state as was physical inquiry a few centuries ago— only "more so". . . . At present we are not even aware of what the *problems* are in and of the human field. On the overt, so-called "practical" side, the ever-accelerating rate at which the troubles and evils, as the raw materials of problems, are piling up and spreading offers a convincing demonstration of backwardness. . . . The assumption is that the various important aspects of organized human behavior are so isolated from one another as to constitute separate and independent subjects, and hence are to be treated each one by methods peculiar to it in its severalty. The subject matter is chopped up and parcelled out into, say, jurisprudence, politics, economics, the fine arts, religion, and morals.[1]

For a nation like the United States such doubt in the sciences of man is a serious matter, because as a nation we have increasingly placed reliance upon science (or possibly "scientifetishism") for the solution of our problems and for aid in achieving what we believe to be the legitimate goals of the good life.

Part of the distrust in social science is perhaps attributable to faulty communication between scientists and laymen. But part is also undoubtedly due to faulty communication between and within the different disciplines and schools of science themselves. A continuous bickering (sometimes phrased in a grandiose manner) over the interpretation of words and concepts absorbs the energies of some few, but conspicuous, social scientists in publication and by word of mouth. Each of the subject fields has sprouted several varieties of currently unreconciled theory. And a good deal of emotion is from time to time generated over opposing points of view that lack either theoretical rationale or empirical content. Such occasional aberrations in the "behavior of the behavior scientists" are healthy in so far as they indicate freedom for untrammeled inquiry. But experience has shown that for best results in science inquiry should be not only untrammeled but also to a certain degree orderly. And there is considerable substance in the observation that the social or behavior sciences could do with a bit more order in their house.

The more serious critics are not interested in factional academic disputes, but are profoundly concerned with the future of man in Western civilization, and they should not be haughtily dismissed by behavior scholars (scientists and humanists) complacently quarreling among themselves in ivory towers. The thinking public in general

[1] John Dewey, "Modern Philosophy," *The Cleavage in Our Culture: Studies in Scientific Humanism in Honor of Max Otto,* Boston: Beacon Press, 1952, pp. 24–26.

looks to the sciences of human relations for reliable answers to the questions of sheer continued human existence in a world of confusions and, beyond this, it expects science and humanistic studies to provide some guidance toward the fuller realization of human potentialities.

Yet, despite the pessimism of those who see nothing but confusion in the sciences dealing with human social life, if one examines carefully the output of serious modern students of man he will inevitably come to the conclusion that they hold in common a considerable body of verified data and agree on many points of theory, even though, in some cases they may not be aware of the fact because of mutually unfamiliar terminologies and frames of reference. What is needed, of course, in the interests of better communication is a concordance of terminology. This is partly a problem in semantics and the general field of symbolization; but also it would help considerably if social scientists could be induced to accept unanimously the generally accepted rules and definitions of plain English, or their native tongue. Likewise needed is a sorting out and ordering of concepts and propositions that would provide order and reason for coordinated and collaborative scientific thinking and research on man's social behavior and its results. But, first of all, perhaps, the followers of the various disciplines need a common focus on what to think about together. What actually *are* their common concerns, and how can men of good will, whatever their disciplinary labels, help each other to share and understand them?

It is to the latter questions that the present book is mainly addressed. Once these matters are somewhat clearer, it is possible that questions of common terminologies, of the construction of cross-disciplinary theoretical systems, and the like, could be settled without too much difficulty.

Organized effort to achieve a sharper focus on the common resources of theory and data among the sciences of social man is not, of course, something new, although a sense of greater urgency has been expressed since the announcement of the unlocking of the atom in 1945. Nevertheless, the fact that considerable activity toward this goal has been going on for some time should raise the possibilities of success of such attempts made in the present or in the near future. We need mention only a few of the more important and currently viable examples.

In the United States the Social Science Research Council since about 1925 has served as a potent stimulator and coordinator of inter-disciplinary work in the social sciences. Its program, supported by some of the leading foundations, has served as an inspiration and model for many other organizations. Its method of administration which involves the official cooperation of the principal professional societies of social science in the country[2] has had the effect of bring-ing many of the leading social scientists together in the Council itself, on special committees, and in work conferences and the like, or-ganized by the Council. Between 1925 and 1951 the Council had awarded 1079 fellowships,[3] not to mention numerous grants-in-aid, support for special projects, and so forth (although not all such monies were granted for strictly interdisciplinary work).

Since the end of World War II two leading international organi-zations, the United Nations and the Organization of American States, have each established, in effect, departments of social science dedi-cated to interdisciplinary focus on human problems and working closely with humanist scholars. The Department of Social Sciences of the United Nations Economic, Scientific and Cultural Organiza-tion (UNESCO) carries on field investigations, promotes international interdisciplinary conferences, and publishes the *International Social Science Bulletin* as well as a number of other periodicals. The Office of Social Sciences of the Department of Cultural Affairs of the Panamerican Union (of the Organization of American States) pub-lishes a bimonthly journal, *Ciencias Sociales,* dedicated to original investigations and to abstracts of the literature of interest to social scientists in the Western Hemisphere, as well as planning and carry-ing on investigations under its own aegis and also in collaboration with qualified groups from the member nations. Thus an institutional framework has been provided in these (and other) organizations for the bridging of both interdisciplinary and international gaps in our understanding of human life.

Another approach to these matters is being currently encouraged by foundations such as the Carnegie Corporation of New York, the Rockefeller Foundation, the General Education Board, the Wenner-Gren Foundation, and the Ford Foundation. It consists essentially in

[2] American Anthropological Association, American Economic Association, American Historical Association, American Political Science Association, Ameri-can Psychological Association, American Sociological Society, American Statisti-cal Association.

[3] *Fellows of the Social Science Research Council, 1925–1951,* New York, 1951.

placing together physically a number of scientists working on man and focusing them on a certain set of problems. Thus by contagion it is hoped that there will develop a meeting of minds and the development of common theory, concepts, and terminology. The Institute of Human Relations at Yale and the Institute for Research in Social Science at the University of North Carolina are examples of this technique which were established previous to World War II, whereas the Laboratory of Social Relations at Harvard, the Institute for Social Research at the University of Michigan, the Bureau of Applied Social Research of Columbia University, and the Social Science Research Center of the University of Minnesota are instances of postwar organizations of this kind. Likewise, a number of "area research" centers and institutes at various institutions in the United States and abroad represent another approach to much the same goal. Also a number of government established or financed research centers or projects have brought scientists of man together; among these mention may be made of the Rand Corporation projects, the Human Resources Research Institute of the Air Force, certain projects of the Office of Naval Research and the Department of the Army, and the Foreign Service Institute of the Department of State. Finally, to mention no other, the Ford Foundation has announced the establishment of the Ford Center for Advanced Study in the Behavioral Sciences; this will begin a five-year period of operation in 1954. These attempts at synthesis by contiguity, as it were, hold much promise for the long run, but up to the present have not made great progress toward theoretical integration outside their own walls (nor sometimes within them). Unless there is fairly systematic and explicit agreement on concepts and interests, there is always the danger that a research or study unit, even though made up of an admirable representation of the various fields interested in human behavior, may break down into a mere collection of specialists who are friendly with one another and who courteously cooperate under the same roof, in the same field camp, or in the same geographical area, without ever achieving scientific collaboration in the fundamental meaning of the word. The program of the Ford Center for Advanced Study promises to give special attention to this problem.

Another step has been taken by such institutions as Harvard with its creation of a Department of Social Relations, involving a sort of polygamous union between social anthropology, social psychology, and sociology, for the purpose of breeding, shall we say, a body of students with a common knowledge and a common point of view at least in the field of social relations. Since the professors have to know what "social relations" are in order to be able to teach them, they must of necessity come to some agreement among themselves. Already, the arrangement has been productive of a considerable body of collaborative work in theory.

The so-called faculty seminar, which often plays an important role in the programs of the afore-mentioned organizations, has also been widely adopted in institutions without special departments or institutes. It typically involves a group of professors, each a specialist in his own field, who have the motivation and the flexibility of mind to sit about a table at regular intervals with the objective of making a concerted effort to reach mutual understanding. Advanced graduate students are admitted as a rule, and their contributions, coming from individuals not yet hardened to disciplinary grooves, are often of great value. Summer interdisciplinary working conferences, often lasting a number of weeks, have also used essentially the same technique and have been sponsored with success by several of the foundations. Something similar, although usually not on so deep or intensive a level, is also involved in the panels of expert consultants being invoked with increasing frequency to deal with problems of human relations by government agencies and business corporations. Joint departments administratively uniting two or more of the behavior disciplines are found in many universities, and "bridging fellowships" enabling students of one field to become familiar with one or more other fields are offered by both universities and foundations.

In the light of these activities and efforts, it is evident that there is a widespread scientific desire to do something more constructive than merely to decry the present uneasy state of the human species and its societies and cultures. However, it is becoming apparent that mere interdisciplinary juxtaposition in itself will not necessarily produce a useful science of social man, any more than locking up some gentlemen of different nationalities in a single room will automatically produce a United Europe. Something more is required than polite social relations between different kinds of experts (although this is conspicuously needed in some quarters), and something more than mere pious hopes. It seems to be clear that we must have a *strategy of a science of social man,* a strategy of development in theory and investigation. Apart from new discoveries and more far reaching theory, if only that which is currently known to scientists dealing with humanity could be brought to focus, mankind as a whole would have less reason to tremble in the face of the "unleashed forces of physical nature."

One obvious way to begin, in the light of present knowledge, has seemed to be to subject the theoretical developments of the "core" human sciences to scrutiny with two objects primarily in mind: (a) to uncover those theoretical elements which, in fact, anthropology, sociology, and psychology hold in common, although such "common denominators" may at times be obscured by cloudiness of verbiage or, even, by the actual unawareness of their proponents of common foci or goals among the several disciplines; (b) to search out those theoretically unformulated convergences of findings and of interest which have so far emerged in the three disciplines and which, once they are better understood across disciplinary lines, might lend themselves to theoretical formulation (and further investigation) as part of a common scientific understanding of social man.

Thus we have not set out in this book to formulate a "final system," but we do aim to be of aid and service to all thinkers and investigators who are endeavoring through their own and the coordinated efforts of others to produce a more integrated comprehension of man in society.

Since our emphasis on the three "core" behavioral disciplines is largely a matter of convenience and historical accident, it should be clear that we hope to be joined as soon as possible by members of other disciplines closely related to the problems of man in the project of deciding upon fundamental questions and agreeing (or agreeing to disagree for valid scientific reasons) upon ways of analyzing them and solving them. Furthermore, our expectations are not confined to the social and psychological sciences. In everything we say in this book the biological aspect of man as a functioning organism in social life is at least implicit—and will often be made explicit. There are, in fact, no serious scientific students of man any more who dare to ignore his "bestial" affiliations. The fact that the human species is a certain type of animal has long been established and recognized. And the limitations as well as the potentialities of performance of this type of organism must be taken into account in any general theory of human activity or accomplishment. We all recognize that a fully rounded theory of human behavior would embody important and fundamental postulates concerning the biological nature of man. In the following discussion, however, we can hardly do other than take this for granted, since the emphasis is upon common theoretical problems of cultural anthropology, psychology, and sociology as at present understood.

To some the phrase "social man" may seem tautological because, on the basis of data, any comprehensive and realistic view of the activities of the human species inevitably records the social context. But we have used this phrase in order to indicate our recognition that from certain points of view it *is* possible to study the human organism and its behavior (in the widest sense) without direct reference, at least, to the social milieu. The anatomist, physiologist, physiological chemist, and other specialists on the organism may with perfect right exclude the social or interactive aspects of human behavior, and in so doing may and do make valuable contributions to our knowledge of the behavior of the human organism *qua* organism. But even the most narrowly specialized among the practitioners of these specialties are aware that their data and theory alone are not sufficient to provide a key to the understanding of human beings as they actually carry on their essential human activities, namely, as social animals. To make a long story short, the available knowledge of the organism as such must be integrated eventually with a vast body of data on the organism as a social being, the creator and follower of cultural patterns, the victim or manipulator of social systems, the originator or dupe of values, and so forth.

For this book the procedure, in simple terms, was as follows. Apart from the editor, two prominent theoretical workers (each has done notable empirical research as well) joined in the enterprise from each of the "core" disciplines involved, namely, anthropology, psychology, and sociology. Then each writer undertook to consider one of the sister disciplines in terms of possible common elements of theory, interest, or findings that in his view would enhance the theoretical scope of his own field. This review was not confined to the more *obvious* common denominators; the emphasis was placed also upon what might be called "viable possibilities" of, shall we say, "interdisciplinary miscegenation."

Although this process has been carried out systematically, no attempt toward complete documentation has been made. Nothing less than a publication of encyclopedic size could begin to approach comprehensive documentation of theoretical notions in our three disciplines as they have appeared, been modified, or superseded during even the past one hundred years, to say nothing of the entire historical period of recorded speculations. A certain amount of documentation is perhaps unavoidable, but in the following pages that amount has been presented in miserly rather than spendthrift fashion. The bibliography at the end consists merely of selected references to works considered especially significant in the present connection.

The authors felt that for a work of this kind at the present time a rigid, formal presentation was to be avoided. Their method of collaboration also evaded the pomposities of academic protocol. After some preliminary exchange of ideas by correspondence the authors met in an intimate week-end conference in New York at the Wenner-Gren Foundation for a thorough face-to-face discussion of the book. While the writing was going on a lively exchange of letters served to keep the participants mutually aware of the development of the various chapters. After the writing had been practically completed, another week-end conference was held in New York, this time at the Social Science Research Council. In the latter discussions the writers, acting as a team, subjected the manuscript to a thorough critical examination, which resulted in some subsequent revision. Thus, although each of the co-authors has his name attached to a particular chapter, I believe that it can be truly said that each chapter and the book as a whole is the product of joint authorship of the entire group to a degree not usually encountered in works of this type. Every page has had the benefit of comments, suggestions, and, in some cases, extensive revisions offered by the co-authors. Nevertheless, once the chapters were written, the editor made no major changes designed to force them into a uniform pattern. It was felt that a variety of approaches would add freshness and spontaneity to the presentation. In our joint discussions a few points arose—apparently relatively minor ones—on which all of the authors were not in complete accord. We decided that these friendly "agreements on disagreement" should not be expurgated, but that, on the contrary, they might prove to be stimulating to readers. Such as they are, they may be occasionally discerned on the following pages.

Insofar as such procedures have succeeded in producing useful results in this book, they may in themselves serve to demonstrate that it is possible for specialists from several disciplines to work together in fruitful collaboration. However, it is, of course, understood that nothing said here necessarily binds any of the authors to maintain his present point of view indefinitely.

CHAPTER 2

Sociology and Anthropology

GEORGE PETER MURDOCK

I

SCIENCE IS A PART of culture. To be sure, it belongs with those aspects, like technology, economy, and warfare, whose changes depend heavily upon success or failure in coping with realities in the external environment rather than with aspects like language, art, and religion, in whose dynamics a substantially greater role is played by internal imperatives and "historical accident." Yet even science exhibits trends, fashions, and differential social support. Its course of development thus conforms at least as closely to the ordinary dynamics of cultural change as to the abstract ideal of a coordinated rational attack upon the unknown.

This is probably particularly true of sciences in their fledgling phase, which sociology and anthropology have barely begun to outgrow. A consideration of the relationships of these two disciplines today and of their probable interdependence tomorrow must therefore take account of their historical development and of the devious ways in which they have interacted in the past.

At the risk of but slight oversimplification, the development of sociology and anthropology can be likened to the life histories of two individuals whose ways have crossed on numerous occasions. The date and exact circumstances of the birth of both are obscure, but not so their parentage. Sociology was clearly born of a union between social reform and philosophy, anthropology of one between antiquarianism and natural history. And the careers of both have given ample evidence of the influence of heredity.

We also know exactly when and how both infants were weaned and thereby established as independent personages. For sociology this event was presided over by Auguste Comte (1798–1857), who was also the child's baptismal godparent. The positive philosophy of Comte was appropriately tempered with a strong ingredient of social meliorism from Saint-Simon (1760–1825). For anthropology the officiant was Adolf Bastian (1826–1905), a colossus whose wide-ranging travels and bulky tomes brought such acclaim that he was able to establish autonomy for the child in whose blood ran both the interest in archeological and folkloristic antiquities and the urge for the systematic collection of artifacts and customs from all corners of the globe.

The socialization of the two children was hardly under way when both came under the influence of a prestigeful but dominating foster parent known to history as Charles Darwin (1809–1882). Identifying strongly with him, they overlearned his doctrine of organic evolution, and for half a century, i.e., until the period of adolescent revolt set in, they reflected his influence in a series of unilinear theories of social and cultural evolution. Some of the giants of this period, like Bachofen, Frazer, Lubbock, McLennan, Morgan, and Tylor, are usually classed as anthropologists; others, like Durkheim, Lippert, Spencer, and Sumner, are more commonly ranked as sociologists. From the perspective of the present, however, they all look remarkably alike, not only in their common use of ethnographic data and a comparative method but also in the majority of the conclusions they reached. From Darwin until after the turn of the century, sociology and cultural anthropology were nearly indistinguishable.

Though the theories of unilineal cultural evolution so widely accepted in the nineteenth century are now almost universally discredited, three of the sociologists of the period have left important and enduring impressions on anthropology, as well as on their own discipline. The first of these is Herbert Spencer (1820–1903). Though it is currently fashionable to discount his *Principles of Sociology,* this work contributed substantially to the definition of the subject matter of anthropology and provided the first outline of what has come to be known as the universal culture pattern, in addition to adding permanent though modest increments to our knowledge of particular aspects of culture, such as religion. Perhaps even more important in the long run has been the influence of his *Descriptive Sociology*. This work, so little known among sociologists that the author has encountered few who have even heard of it, inaugurated a commendable effort to organize and classify systematically the cultural data on all the peoples of the world for the advancement of cross-cultural research, and thus clearly foreshadowed the development of the present Human Relations Area Files.

The second of the significant early sociologists is Emile Durkheim (1858–1917). Though he accepted the unilinear evolutionary theories current in his day, he did not emphasize them. He also drew heavily upon ethnographic materials, but instead of amassing quantities of cases torn from their cultural context he immersed himself in the literature on one group of primitive peoples, the Australian aborigines, and sought explanations of such cultural phenomena as religion in terms of their functions in social interaction. He thus paved the way to the development of functionalism, which ultimately infused new vigor in anthropology after both evolutionism and historicism had revealed their essential sterility. Malinowski, Radcliffe-Brown, and Redfield, and through them their many students, were strongly influenced by Durkheim in diverse ways, and the structural approach now dominant in British social anthropology is specifically his intellectual grandchild.

Unfortunately Durkheim, in his effort to stress the autonomy of cultural science, erroneously rejected psychology as irrelevant. Consequently, though the integration of sociology and anthropology has indubitably been advanced through his influence, the rapprochement of both with psychology has been inhibited. Those anthropologists who have followed his lead, for example, have remained aloof from the promising developments in behavior theory and in "culture and personality." Many have tended to short-circuit genuine scientific problems by reifying social structure, much as other anthropologists have reified culture. Either approach would deny psychology a place in a science of social or cultural behavior.

The third influential sociologist of the early period, William Graham Sumner (1840–1910), foreshadowed functionalism in another way. He insisted on what is now a commonplace of anthropological thought but was novel in his day, namely, that culture is adaptive, satisfying individual and societal needs and altering over time in response to the changing conditions of life. He was the first to promulgate explicitly the doctrine of cultural relativity, shocking his contemporaries by asserting that even slavery, cannibalism, and infanticide are adaptive and socially justifiable in the societies that normally practice them. In his *Folkways* Sumner also demonstrated that the elements of culture are very much more than neutral "traits" or shared habits, being commonly invested by the members of a society with affect, moral values, and an awareness of sanctions. These views are thoroughly congruent with the participation of psychology in cultural science. Though Sumner's direct influence on anthropology has been slighter than that of Durkheim, his indirect impact has been enormous, and he probably stands closer than any other social scientist of the late nineteenth century to the theoretical position collectively represented by the authors of the present volume.

The years just preceding and following 1900 brought the first serious doubts as to the validity of the unilinear evolutionary theories which had bound sociology and anthropology in a premature united front. The accumulation of ethnographic evidence began to reveal the extreme importance of cultural diffusion, undermining a basic assumption of evolutionism, and sociologists like Howard and Westermarck suggested alternative explanations for customs which had previously been accepted as survivals of earlier stages of social development. The basis of their unity shattered, the two disciplines parted company, and for several decades followed courses reflecting their separate pre-Darwinian origins.

Sociologists, with a few notable exceptions like Keller and Thomas, lost interest in ethnography and abandoned comparative studies. Many reverted to the well-trodden paths of philosophy— some, like Ward, Park and Burgess, and Sorokin to build ambitious theoretical systems; others, like Giddings, Cooley, and MacIver, to speculate in the armchair on a lesser scale, too often with the comforting but enervating assumption that the study of human behavior, as a *Geisteswissenschaft,* can never hope to make successful use of the methods of science. Another large group returned to a frank interest in practical social problems—delinquency, dependency, divorce, racial discrimination, population, etc. Not being averse to observation or obsessed with the greater virtue of theorizing, they did amass a modest body of verified knowledge. But since they naturally had little interest in data from societies other than their own, they would have converted sociology into a culture-bound discipline, like economics and political science, had it not been for the system-builders, who fought valiantly for a science of universal validity comparable to anthropology and psychology. Ultimately, of course, the fact-mindedness of the one group and the theoretical preoccupation of the other were to fuse and produce a mature discipline which could make common cause with the sister sciences of man.

In the meantime anthropology, too, was exhibiting signs of regression. In part its practitioners stressed the original natural-history objective and strove energetically, through intensive field work, to record in detail the customs of the simpler peoples while there was yet time to recover them. In part they followed the equally time-honored antiquarian lure, unearthing the remains of the more spectacular prehistoric civilizations, following hobbies in primitive art and folklore, and tracing the origin and geographical spread of interesting traits. Concern with the study of man as a science gave way to a widespread insistence that the only legitimate objective of anthropology is the reconstruction of history for the periods and regions lacking written records. Theories of prehistory involving the intercontinental diffusion of great culture complexes were elaborated by Smith in Great Britain, Graebner in Germany, and Schmidt in Austria. American anthropologists were in general more cautious, preferring to build the edifice of prehistory more slowly, brick by brick. The formulation and testing of scientific hypotheses, especially by comparative methods, was considered, if not entirely taboo, at least premature and thus to be deferred to some millennium when the ethnographic evidence would be complete.

The most striking characteristic of American anthropology during the first quarter of the twentieth century, which was dominated by Franz Boas, was its negative attitude toward most theories which had gained a measure of acceptance in the human sciences. It combatted, nearly always with eventual success, not only unilinear evolutionism, but also geographic and economic determinism, the organismic analogy, the group mind, racism, instinctivism, eugenics, and theories of primitive mentality. In short, it swept the decks clean, retaining little except the basic concept of culture itself. In so doing, it rendered a genuine service to the human sciences, as is apparent today from our longer perspective, for nearly all that went overboard is now recognized as rubbish which needed to be swept away before genuine scientific advance was possible.

During this period sociology and anthropology almost completely lost contact. The ambitious philosophers and earnest meliorists who dominated the former could hardly be expected to be impressed with the importance of the purely historical study of grubby and vanishing peoples in remote places. The fact-minded and field-minded anthropologists in their turn viewed with scarcely veiled contempt the "arm-chair theorists" and "do-gooders" in the other camp. This had the unfortunate effect of closing their eyes to the important gains, however few these may have been, that were actually being achieved in sociology. The field of culture theory itself, presumably the special preserve of the anthropologist, provides a spectacular example. Substantial contributions to the dynamics of culture change made by the sociologists Keller and Ogburn passed unnoticed in anthropology, which continued to limp along for another quarter century with crude concepts of invention and diffusion, derived from the nineteenth century sociologist Tarde and only inconsiderably improved, before it achieved a comparable level of theoretical sophistication.

By the 1920's Boasian anthropology had exorcised all its devils and, being committed to a negative bias, was threatened with the sterility that ultimately overtakes all "schools." It was rescued by a revival of interest in scientific problems, which stemmed from several sources. Wissler, who had been trained in psychology, promoted the study of acculturation. Sapir and M. Mead, by different paths, became interested in depth psychology and pioneered the field of "culture and personality" which was to lead to a rapprochement with psychology. Still more electrifying was the impact of "functionalism" as promulgated in different ways by Malinowski and Radcliffe-Brown though derived ultimately, as previously noted, from the sociologist Durkheim. Fortunately functionalism was never accepted as a dogma in American anthropology but rather acted as a ferment, leading to an integration of synchronic with diachronic interests, methods, and objectives and to an eclectic, dynamic, and basically scientific approach well exemplified, for instance, in the work of Linton and Eggan.

It is an ironic fact that most of what modern anthropology has absorbed from the functionalists it could have learned decades earlier from American sociology had it chosen to listen. Essentially the lesson from functionalism has been that cultures and sub-cultures are organically related to the structured social groups and sub-groups that carry them, which has been axiomatic in sociology at least since the days of Sumner. Utterly incredible as it must seem to the psychologist and sociologist, it is nevertheless almost literally true that no work by an American anthropologist recognized this fact until the appearance of Linton's *Study of Man* in 1936. The prevailing point of view was expressed as late as 1937 by Lowie in his *History of Ethnological Theory*, where he expressly rejects the position of Malinowski and insists (pp. 235–236) that "a culture is invariably an artificial unit segregated for purposes of expediency" and that there is "only one natural unit for the ethnologist—the culture of all humanity at all periods and in all places"! The present writer cannot conceive of a more telling object lesson for the need of interlearning among the sciences concerned with human behavior.

II

Happily, evidence is increasing that anthropology and sociology, as well as psychology, have overcome the traumas of their birth and weaning, corrected their childhood dependency on biology, and resolved their adolescent conflicts, so that they are prepared to interact with one another on an adult level. It is clear that sociology and anthropology are becoming reconciled after their long separation and already share, to a marked extent, the same body of basic scientific theory. Their further rapprochement with one another and with psychology, however, will be hindered rather than promoted by failure to recognize certain fundamental differences between them, stemming largely from their historically divergent developments. In attempting to clarify these differences, the writer must not be understood as wishing to justify or perpetuate them, or to stake out jurisdictional claims for either discipline, or to make points in favor of one at the expense of the other. He happens to be a practitioner of the one who had his professional training in the other, and he acknowledges a loyalty to both. His aim is solely to set forth their major differences, as he sees them, in the belief that this may point the way to the areas in which each can learn most from the other in the future.

Anthropology owes one of its major differences from sociology to its natural-history antecedents. The natural sciences, such as botany, zoology, and geology, recognize a professional obligation, over and above the investigation of theoretical problems, to record systematically the forms of plants and animals and natural formations in all the parts of the earth which they visit. Anthropology shares this sense of obligation. Field work is the *sine qua non* of professional standing, and the ethnographer, whatever special problems he may go to the field to investigate, is expected to bring back and publish, not only an answer to his special problem, but also a descriptive account, as complete as he can make it, of the entire culture of the people studied. The result has been the accumulation of a vast body of professionally gathered descriptive materials on the approximately 3,000 peoples of the primitive world.

Among the other human disciplines only geography shares this natural-science sense of obligation for full descriptive reporting. The psychological sciences, by contrast, have produced only a handful of complete case studies of individuals, their presumed special subject of investigation. In sociology, "sociography" enjoys nothing like the standing of ethnography in anthropology, and the body of accumulated descriptive studies produced by professional sociologists is inconsiderable, consisting mainly of a few regional descriptions, a few accounts of urban groups or areas, and a few community studies, of which a substantial proportion have actually been produced by anthropologists like Arensberg, Davis, Gardner, Miner, Powdermaker, Warner, and West.

Much sociological research starts with a scientific problem, for which the researcher gathers the material he needs by questionnaire, schedule, interview, or other techniques; he then publishes so much of the data as shed light on the problem, but no more. Most other sociological research utilizes data gathered by official agencies, notably census reports and vital statistics for the Registration Area. Sociologists display laudable ingenuity in inducing other people, like the War Department or the Census Bureau, to assemble the information they require, but remarkably little motivation to gather and report descriptive data through their own efforts.

An ironic consequence of this difference is that there is scarcely a single aspect of modern American culture, not alone for the country as a whole but even for any single community within it, which is as adequately described in the literature as is the comparable aspect in scores or hundreds of primitive societies. That this is not due to population size or cultural complexity is demonstrated by numerous adequate descriptions of complex African societies with populations numbering hundreds of thousands or often millions, as well as by the rich sociographic literature on several Eastern European countries.

The fields of research of sociology and anthropology are, of course, differentiated, the one working mainly in complex Western civilizations, the other chiefly among simpler preliterate societies. This is, however, merely a practical division of labor, and is in itself of little consequence. Practitioners of both disciplines have amply demonstrated that each can operate effectively in the other's field. There is no "anthropological method" or "sociological method" as such, characterized by special virtues. Nevertheless, the division of labor between the two disciplines has given rise to a fundamental difference in the kinds of data primarily sought and in the techniques for securing them—a difference which, because largely unrecognized, has been the chief source of friction between the two disciplines.

Anyone who studies a society with a culture differing markedly from his own—whether he be an anthropologist, a missionary, a colonial administrator, or a sociologist—must first gain a comprehension of the patterned norms of that culture before he can even begin to understand the behavior of the people. And if he writes up his findings for members of his own society, as he usually does, his first task is to make those norms explicit. The inevitable result has been that anthropologists have devoted their primary attention to patterned behavior, i.e., to those norms which are verbalized as the ideals to which behavior should conform, are taught to each oncoming generation, and are enforced by the formal and informal punishment of deviations. Any member of the society knows a large proportion of these norms, whether or not they govern his own behavior, for the norms incumbent upon persons in other statuses constitute an aspect of his expectations in his social interaction with them. To gain an accurate account of patterned behavior, therefore, an ethnographer needs only a few competent informants, selected judiciously but not necessarily in accordance with any standard sampling technique.

The situation is quite otherwise when a scientist is studying his own society and is writing up his results for other members of the same society. He and his readers, as participants, already know the major norms of the culture, and it would be trite to detail them. What neither he nor they do know, however, is the incidence of unpatterned behavior—both variations within the limits of prescribed patterns and deviations from such patterns. And these, rather than the norms themselves, are what he reports. American "voting habits," to an anthropologist from a different culture, would mean, first of all, such patterned practices as being transported to a polling place in a car provided by a political party's headquarters, checking with the registrars, and X-ing a ballot or pulling a lever on a voting machine; to an American sociologist, on the contrary, the term refers exclusively to such non-patterned behavior as how many people turn out to vote and how many vote Republican and how many Democratic. An accurate record of the incidence of unpatterned behavior is obviously impossible, short of a full count, except through the use of the most refined sampling techniques.

It should now be clear why sociological community studies, like those of the Lynds on Middletown, though commonly compared with ethnographic reports by anthropologists, are in fact poles apart. They include data of a basically different type, with surprisingly little overlap. It should also be apparent why we know so little about American culture. We know so little precisely because, as participants, we know so much to begin with, and are thus almost inevitably motivated to concentrate on the enormous mass of behavior about which we know little or nothing, namely, unpatterned behavior.

To students of man, of course, all behavior is important, patterned or otherwise. It is a serious defect of anthropology that it has traditionally concerned itself so exclusively with the former. Sociology has made a major contribution in calling attention to the importance of the latter. That anthropology is learning this lesson is revealed by the increasing attention being paid in recent ethnographic literature to variations in behavior within the prescribed patterns and to deviant behavior. If the anthropologist in the future provides the sociologist with more data of a type which he can compare with his own, perhaps the latter will reciprocate by telling the anthropologist some of the things he would like to know about American culture.

Doubtless because of its almost exclusive concern with patterned behavior, cultural anthropology possesses a much more unified body of theory than does sociology. This revolves about the concept of culture and embraces, among its major divisions, (1) culture dynamics, or the body of theory concerning the processes by which culture changes over time, (2) social structure, or the body of theory concerning the patterning of interpersonal relationships in social groups and status categories, and (3) culture and personality, or the body of theory concerning the transmission of culture and formation of personality norms through the socialization process. In addition, of course, there are special segments of theory concerned with particular aspects of culture such as language, technology, art, religion, and government. To a remarkable extent this system of cultural theory is shared by all anthropologists. There are, of course, still differences of emphasis, and even some basic cleavages, but they are notably fewer and less critical than they were a quarter century ago.

Impressive as it is in its essential unity, this body of theory is by no means the exclusive creation of anthropologists. Culture and personality derives mainly from psychology, and social structure to a considerable extent from sociology. Even the central concept of culture itself owes as much to the sociologist Sumner as to any anthropologist who has ever lived, and lesser contributions of significance from sociologists are legion.

Sociological theory—at least to the extent that it is not derived from or shared with psychology or anthropology—is a much less monolithic structure. It gives the impression, if one judges by the periodical literature rather than the tomes of the system-builders, of being composed of a very large number of fragmentary and isolated propositions, each tested and at least tentatively validated, which are mainly of a relatively low order of generality and often hardly more than empirical generalizations.

Several possible explanations come to mind. One is that the very wide range of behavior encompassed presents a problem of much greater complexity than has faced the anthropologist in his concentration on patterned behavior. A second flows from the fact that no conclusion from a piece of research conducted in one society, however promising the hypothesis and convincing the proof, can be accepted as an increment to the general science of human behavior until it has been cross-culturally validated. Until then it can be assumed at best to be established for the one society only, and there is no way of knowing to what extent the seeming confirmation is due to the common nature of man or the universal conditions of social life and to what extent it may merely reflect conditions peculiar to the particular society or its unique historical assemblage of culture traits. Wherever factors of the latter general type preponderate, of course, verified propositions that are unrelated and even occasionally unreconcilable are to be expected. Since exceedingly few of the propositions established by sociological research have as yet been cross-culturally validated, their seeming heterogeneity may simply be a result of the fact that they represent an as yet unassorted mixture of intra-cultural or particularistic generalizations with cross-cultural or universally valid generalizations.

A third possible explanation of the heterogeneity of sociological as contrasted with anthropological theory may reside in the differential determination of research problems in the two fields. In sociology research attention is commonly directed to areas of social problems, i.e., to situations defined as unfortunate, particularly where common-sense efforts at correction have failed and more precise knowledge is clearly called for. The melioristic tradition in sociology makes such areas congenial to many, but probably even more important is the availability of large research funds for their investigation. Industries wish to reduce labor turnover and increase production; the armed forces wish to improve morale and select effectively for specialized training; the media of mass communication wish to sell products or influence attitudes more effectively; public opinion deplores crime, high divorce rates, poor housing, and racial discrimination; and organizational, foundation, and government funds are made available for research that may lead to their reduction. The demand for trained sociologists from such sources is imperative, and those who respond need not work on a shoestring.

Among anthropologists, by contrast, the problems for investigation are much less frequently set by others. Funds for field research are restricted by geographical area, if at all, and the worker is relatively free to select his own scientific problem. If he goes to the Bongo-Bongo, neither the foundation that backs him nor anyone else is so concerned with the high divorce rate or poor housing of the tribe as to direct his research to a particular subject. He is at liberty to concentrate on whatever he pleases, and he commonly picks a topic, like kinship, which is no social problem at all. With this freedom, the determining factor in his choice is likely to be the theoretical structure of his discipline; he chooses a specific research project which will revise that structure or advance one of its frontiers. In consequence, anthropological research tends constantly to expand the discipline's central core of theory rather than to proliferate a heterogeneous body of uncoordinated special theories.

The sociologist's difficulty is not only that his research tasks are so commonly set by others, rather than by himself, but also that he is thereby compelled to start with a situation which needs explaining instead of an abstract question which needs an answer. This is, of course, essentially the old problem of pure versus applied science—

a dichotomy which the writer considers in many respects misleading and thus prefers to restate. The crucial distinction seems to be
that a scientist who starts with a theoretical question, a "pure-sci-
ence" hypothesis, can select a situation adapted to testing it, one
in which he can exert maximal control over the variables, whether
in the laboratory or in the field, whereas the scientist who begins
with a situation cannot expect comparable control of the variables,
must be prepared for extensive exploratory work to isolate them, and
is under compulsion to assess the influence of every important variable, not merely that of a particular one manipulated to test a single
initial hypothesis. Since the variables in any situation are likely to be
numerous and diverse, not only the sociologist but any scientist
whose research begins with situations will almost inevitably emerge
with a number of validated hypotheses, and these will often be dis-
crete, unrelated systematically to one another or to those established
by workers in different fields, and of a relatively low order of generality.

   That the total body of verified theory which has resulted from
sociological research compares unfavorably with that of anthropology in unity, in generality of application, in sheer quantity, and in
potential significance for the integrated human science of the future
would probably be admitted by most sociologists, and is certainly understandable in view of the special difficulties encountered. What
is less generally recognized is that certain positive advantages have
accrued from this experience which anthropology does not share and
urgently needs to acquire. The very handicaps faced by the sociologist—confinement to a single society without the opportunity for
cross-cultural validation, selection of his research areas by others
than himself, and the compulsion to assess the numerous variables in
complex situations—have exerted a highly salutary disciplinary effect. They have developed in him a high degree of awareness of the
nature of scientific method, a respect for science comparable to that
prevailing in the natural sciences, and an amazing ingenuity in adapting scientific means to scientific ends.

   By comparison, anthropologists are extraordinarily naïve in scientific matters. Many frankly confess a humanistic rather than a
scientific orientation, and not a few are openly anti-scientific. Among
those who are actually engaged in research problems which can be
classed as scientific, only a handful are adequately grounded in scientific method, and many of these are committed to one method and

are skeptical of others. Those who, like Kluckhohn, are both genuinely sophisticated and broadly oriented are rare indeed. In sociology, men with a flaming zeal for science are not uncommon. The writer thinks instantaneously of such diverse figures as Keller, Lundberg, and Stouffer, but he has never encountered such a man in anthropology. The low status of scientific interest and awareness in anthropology is assuredly the most serious handicap to the full participation of this discipline in the integrated human science of the future. Here, if anywhere, there is room for learning from both sociology and psychology.

The scientific superiority which sociology enjoys over anthropology is perhaps most obvious in the field of methodology. Anthropologists have, indeed, evolved the wholly admirable "genealogical method" for field research, which even sociologists could profitably adopt, but in general they have been uninventive. A striking example is provided by research in culture and personality with its almost slavish dependence on projective tests borrowed from psychiatry. When it comes to the formulation and testing of hypotheses, anthropologists reveal little comprehension of the requisites of a viable scientific theory and even less of the methods which science has devised for putting such a theory to the test.

The misunderstanding of scientific method is perhaps most extreme in that group of anthropologists which makes the most vociferous pretensions to being scientific and comparative—the British structuralists headed by Radcliffe-Brown. The alleged "laws" of this school turn out, upon examination, to be verbal statements like "the equivalence of brothers" or "the necessity for social integration" which fail completely to specify the concomitant behavior of variables, and their proof invariably rests on the inspection of a single society with perhaps some reference to a small and utterly unrepresentative sample of other societies in British colonial territory. American anthropologists tend less to distort than to ignore the canons of science. They too have made excessive use of that crudest of all scientific methods, the "clinical method" in the broadest sense, which consists essentially in the intensive examination of single complex cases and depends on skill and experience alone to assess the influence of different variables. Little use has as yet been made of such relatively refined methods as experiment (not always impossible with human subjects), analysis of the crucial case (one selected because all variables occur naturally in the combination desired), and statistics (with its varied techniques for establishing covariance in large numbers of cases).

Even the most casual contact with sociologists, or with the literature they produce, reveals their superior scientific sophistication. Anthropologists cannot afford to rest on their laurels, contemplating with complacency their own impressive structure of culture theory and smugly treating the methodological efforts by which their colleagues in sociology attempt to cope with peculiarly difficult problems as a slightly ridiculous game of mental gymnastics. Even granted that these efforts are occasionally excessive, have the anthropologists been guilty of fewer excesses when they have faced equally knotty problems, as in trying to account for national character? If anthropology is to become a science, and to play an equal role with its sister disciplines in building the unified human science of the future, it has much to learn about science and scientific method from sociology.

III

We have now reviewed the history of the relationships between sociology and anthropology, assessed the respective strengths and weaknesses of each in relation to their differential problems and development, shown some of the more important contributions which sociology has made to anthropology in the past, and indicated some of the genuine achievements of the former which the latter has not yet adopted or taken full advantage of. It may not be amiss, in conclusion, to assume the role of prophet and, by projecting these trends into the future, attempt to visualize what may be the distinctive contributions of the two disciplines to a unified science of human behavior.

In the first instance, the relations of both to psychology will require some revision. Each will have to become reconciled to the near certainty that the basic mechanisms of behavior will be established primarily by psychological research rather than by their own efforts. Concepts like "the processes of social interaction," "the processes of cultural change," and "the socialization process" must be recognized as nothing more than psychological processes, such as those of perception, learning, and personality development, operating under the special conditions created by human social life. There must be no more reification of "culture" or "social structure" as causal forces, no more assumption of a special superindividual or superorganic level of phenomena characterized by a body of principles inaccessible to the psychologist.

Psychologists, in their turn, must desist from their efforts to explain social and cultural phenomena in terms of behavior mechanisms alone. All such attempts in the past have failed dismally, and they will fail equally in the future, for mechanisms, whether of learning or of personality, produce differential results depending upon the particular conditions of material environment, social organization, and culture prevailing in a given situation. It is the special province of the sociologist and the anthropologist to study such conditions and to determine what constellations thereof, in conjunction with behavioral mechanisms, produce this or that social or cultural manifestation. It is not enough that either side should admit the importance of the other's contribution. Both must recognize that each holds one indispensable key and lacks another, and that few if any of the rooms in the future mansion of human science can be unlocked without both.

The relations between anthropology and sociology promise to be somewhat different. The most crucial determinant would appear to be the fact that one holds most of the resources, the other most of the tools with which to exploit them. Anthropology has at its disposal, in the riches of ethnography, evidence concerning an immensely wider range of variation in human behavior than has any other discipline. It has access to thousands of nature-made experiments, combining conditions of social and cultural life that are utterly impossible to reproduce artificially. They provide the ideal ultimate testing ground for theories of human behavior. The human scientist of the future will certainly look with equal skepticism upon statements alleged to be generally valid for man when made on the basis of animal experiments unchecked with human subjects and when based on research in a single society unverified by a cross-cultural test. To us as citizens, of course, our own society looms as enormously important; as scientists, however, it remains an insignificant single case among thousands, and statements concerning it, no matter how firmly buttressed, can no more be accepted as true of mankind in general than could psychiatric observations on a single patient.

If anthropology appears qualified and even destined, by virtue of its superior resources, to become the final proving ground of behavior theory in the future, it sadly lacks the means to realize this potentiality. These means, among which by far the most important are sophistication in scientific theory and versatility in scientific method, are found in abundance in the sister discipline of sociology, where they lie for the most part unnoticed. Their utilization must follow one of two alternative courses. Either anthropology must go to school and learn thoroughly the skills which sociology has to offer, thereby equipping itself to perform the task, or sociologists will revive their nineteenth-century interest in ethnography, apply to it their wealth of new techniques, and succeed where Spencer, Durkheim, and Sumner failed in establishing a genuine cross-cultural science of man. The Human Relations Area Files would incomparably facilitate the latter.

Whatever the outcome, sociology will certainly retain one exceedingly important function. Though the general validity of theories about human behavior must depend ultimately upon cross-cultural checking, our own society offers obvious advantages as a preliminary testing ground. The costs in time and funds are minimal. The researcher commands the language of his subjects. As a participant, he knows their culture intimately, and can adapt his techniques with precision to the social situation as well as to the requirements of the problem. He can draw upon enormous resources of data gathered by others to augment those assembled by himself. There are, to be sure, certain scientific problems which can be tackled only in another cultural setting. For the great majority, however, our own society provides the ideal scene for exploratory work, for experimentation with novel techniques, for the formulation and revision of hypotheses, and for their thorough initial testing.

Anthropology, if it rises to the occasion, may ultimately become the final arbiter of the universality of social-science propositions and may even contribute, next to psychology, the largest share of really basic theory. Sociology, however, gives promise of greater creativity in respect to theories in the lower and middle ranges, and certainly in regard to methodology, and it is quite conceivable that in some areas, like that of small group research, it may yield scientific principles of substantial magnitude.

Developments in all three disciplines over the past quarter century give numerous indications that psychology, sociology, and anthropology are gradually merging their most important individual contributions to form what may some day become a unified science of human behavior. Serious obstacles remain to be overcome in each of the fields, and this writer has taken pains not to minimize them. Nevertheless, he is heartened by the fact that those who share the vision of ultimate common achievement, typified by his co-authors in this volume, include a very large proportion of those to whom their respective professional colleagues are accustomed to turn for leadership.

CHAPTER 3

Anthropology and Psychology

M. BREWSTER SMITH

To TREAT THE interrelations of anthropology and psychology from the point of view of psychology, as falls to my part in this joint undertaking, should presuppose agreement on the meaning and scope of the terms being related. Yet the state of affairs that makes this venture desirable includes, I think, the fact that the boundaries dividing these disciplines from one another and from sociology are by no means sharp, stable, or justified on obviously rational grounds. A source of fruitless jurisdictional disputes and no little confusion, it also gives rise to the hope of attaining a more comprehensive science of social man.

For a starting point, it is probably best to accept psychology, anthropology, and sociology for the cultural and historical facts that they are: academic specializations that emerged late in the nineteenth century, each blessed today with a distinctive body of literature and shared informal lore—a tradition transmitted mainly in college and university departments through relatively standardized training experiences. Joint departments, to be sure, are a frequent occurrence, and an occasional individual changes his affiliation. But at a time when it would be hard indeed to give satisfactory conceptual definitions that would unmistakably distinguish the three disciplines for the uninitiated, practically no one who is professionally involved in them has any doubt about whether he is a psychologist, an anthropologist, or a sociologist. The only adequate way to define each discipline is to point to its history, and to the current activities of the people who agree on identifying one another as members.

Each discipline has, of course, its subdivisions or areas of specialization. Some of these are most loosely interrelated. Consider, for example, the loose bonds among archaeology, linguistics, cultural anthropology, and physical anthropology on the one hand, and constitutional psychology, personality study, physiological psychology, and psychometrics on the other. All of these specialties and many more have their essential contribution to make to the integral study of man. To make my present task feasible, however, I will focus on what has come to be known as cultural or social anthropology, with respect to its contribution to general psychology, and, particularly, to social psychology and the study of personality. I am thus concerned with anthropology and psychology as they bear on the social nature of man, rather than on its biological basis.

While it may be adequate to define a discipline by a pointing operation, it is quite unsatisfying. One looks rather, if not for boundaries, at least for a distinctive theoretical perspective, for a conceptual center of gravity. In the case of psychology, what runs through an otherwise heterogeneous history is a pervading focus on the individual. From beginnings in which interest centered on his experience or consciousness, through a brash adolescence in which only his most objectively observable activities were deemed suitable for scientific study, to the present more catholic outlook in which psychologists resort more readily to unobservable constructs and inferential concepts in their interpretation of behavior, the individual has been the primary reference point.

This has not meant that psychologists have always thought of themselves as studying individuals. Often enough, the person has been lost to sight, whether in the early concern with the description of the "generalized human mind," or in the later pursuit of general principles of behavior in studies of the lowly rat. Modern social psychology, moreover, is struggling to attain ways of dealing more directly with the *interactions* of persons, finding too rigid a focus on the individual as an isolated entity a stumbling block in the way of attaining adequate theory. Nevertheless, descriptions of the organization of consciousness and the principles of behavior, whether they dealt with learning or motivation or social interaction, have with few exceptions been conceived as applying at the level of the individual.

The aspects of anthropology with which we are presently concerned, on the other hand, have had a different focus. From the outmoded phases of evolutionism and diffusionism to the current interest in a theoretical account of culture and social structure, and through the persisting strain of descriptive ethnology, anthropology has come to direct its attention on the organized relations of the members of a society, and on the cultural tradition in which these are imbedded.

The role of the individual in this pattern of interest has been controversial. Except when the anthropologist is working with artifacts of "material culture," to be sure, he secures his data from individuals, either as informants or as actors in the drama of society. After serving as a means in the generation of data, however, the individual traditionally dropped out of the picture, in favor of cultural or social patterns described so as to be relatively independent of the particular persons who happen to carry or enact them.

At the extreme, this tendency has been given theoretical status in the position, taken for example by Leslie White[1], that culture exists at a level utterly independent of the individuals who carry it, with special laws of its own. More characteristic of the current temper, however, is the opposing trend, given its initial impetus by Sapir,[2] to look to the individual for a more differentiated version of the cultural process and for a leverage point at which the theorist can come to grips with cultural continuity and change. Once the individual comes within the range of attention, a host of fascinating problems arise that may carry the anthropologist far into the territory traditionally staked out by the psychologist. But the center of gravity of the anthropologist's concern with the individual, if such can be distinguished, is certainly the interpretation of cultural and social processes supra- or interindividual in scope. No longer disembodied, culture is still the primary focus.

My task of examining the contribution of anthropology to psychology is complicated, however, by the fact that it is difficult to distinguish anthropology from sociology with respect to theoretical focus or principal reference point. It would be convenient if anthropology could be identified as the science of culture[3] and sociology as

[1] Cf. Leslie A. White, *The Science of Culture*, New York: Farrar, Strauss, 1949.

[2] Cf. Edward Sapir, "The Emergence of the Concept of Personality in a Study of Cultures," *Journal of Social Psychology*, 1934, 5, 408–415.

[3] I find Linton's definition as useful as any and will try to abide by it: "A culture is the configuration of learned behavior and results of behavior whose component elements are shared and transmitted by the members of a particular society." Cf. Ralph Linton, *The Cultural Background of Personality*, New York: Appleton-Century-Crofts, 1945, p. 32.

the science of social structure or social systems—of the patterned interrelations of members of a society. But as we all know and Murdock has reminded us, anthropologists have paid major attention to social structure with considerable substantive results. The sociologist, on his part, has implicitly studied culture, even when he has held to general focus on social structure, which, after all, has major cultural components. The distinctions between sociology and social and cultural anthropology, foremost among them the traditional anthropological emphasis on the comparative study of non-literate societies, are historical accidents. Any survey of the contributions of anthropology must therefore treat certain concepts which are also the common property of sociologists.

How Psychology Has Drawn on Anthropology: An Overview

A detailed historical account of the interplay between anthropology and psychology should go back at least as far as Wundt, whose program to study the higher mental processes through their objectified cultural products culminated in a *Völkerpsychologie* that drew heavily on ethnographic materials. But this proved to be somewhat of a by-path in the development of both disciplines. We can more profitably begin by inquiring about some general features of the processes of diffusion and acculturation by which the influence of anthropology has been felt in psychology during the more recent period that includes the present situation and its immediate antecedents.

In cases of culture contact, the anthropologists tell us, items of culture content are more likely to be transferred than the broader configurations that give them meaning in their original setting. So it has been in the contact of psychology with anthropology. With the exception of the broad concept of culture, the bizarre data brought back by the anthropologist from afield have had more impact on psychology than the theories with which anthropologists approached their data. It is the rare psychologist who lacks passing acquaintance with the Arapesh, the Trobrianders, the Kwakiutl, the Zuni, or the Navaho, though a few stock tribes perhaps suffice to establish for him and his students the point that the condition of man is exceedingly various. Much less frequent is the psychologist who has exposed himself to sophisticated analysis of the culture concept, or who has re-

sorted to anthropological descriptions for help in specifying the social environments in which behavior takes place.

This tendency of the psychologist to borrow data rather than theory from the anthropologist is doubtless in no small part the result
of the traditional stress in anthropology on sheer descriptive ethnography. The anthropologist had access to the natural laboratory provided by the wide range of variation among the vanishing non-West-
ern societies; he was preoccupied with capturing a record of these
societies while they were still accessible. Until anthropological theorizing in recent years itself took a psychological direction, the psychologist found the ethnographic data much more interesting than
the anthropologist's speculations about his still meager conceptual
framework. And if the psychologist made only superficial use of the
riches of anthropological data, he was abetted by the stress on a
somewhat monolithic concept of culture in the anthropological writings that diffused to him. One does not have to look very closely at
the facts of very many societies to appreciate the importance of culture; it takes a closer look, to be sure, to see that announcing this
concept only poses a host of problems which it does not solve.

With the development within anthropology of a strong psychological current in the study of "culture and personality," showing
closer affinities to psychoanalytic theory than to the prevalent academic psychology, another line of influence was established. The
popular works in this vein—those by Margaret Mead are the most
salient example—are probably still the most frequent avenue of psychological acquaintance with anthropology, as witness the names of
stock tribes that immediately come to mind to a psychologist. Although the psychological approach to anthropology met with initial
unfavorable criticism in the psychological camp as it did in the anthropological, it may well be that this feedback of modified psychoanalytic ideas to psychologists has something to do with the more important role of psychodynamic conceptions in contemporary psychology. As McDougall foresaw in writing his *Introduction to Social
Psychology*,[4] the social sciences need from psychology an adequate
theory of motivation. The borrowings of anthropologists from psychoanalytic motivational theory have at least dramatized to the psychologist the need to get his own house in order in this conceptual area.
Paradoxically, the seriousness with which psychological attention is

[4] London: Methuen, 1908.

now focused on a re-examination of Freudian psychodynamics may in part represent an anthropological influence on psychology.

In this recent interplay of anthropology and psychology, habits of research and thinking established in the characteristic training patterns of each discipline have influenced the developing channels of communication and borrowing. Anthropological field training has tended to emphasize keen observation, the shrewd use of informants, and naturalistic qualitative description. So long as practical exigencies confront one or two field workers with the task of describing an entire culture on the basis of a year of residence more or less, the field worker is forced into bold extrapolations if he aspires to any synthetic characterization of the culture as a whole. Anthropological methods of collecting and digesting data have therefore had more in common with the procedures of clinical psychiatry than with the methodological nicety of experimental psychology. While the anthropologist, like the psychiatrist, is moved by urgencies in the real situation to glean as much insight as possible from methods that are open to challenge from the purist, the psychologist from the beginning has been ready to risk triviality for the sake of the scientific rigor that is his badge of distinction from philosophy and anecdotal or armchair speculation.

Anthropology, therefore, has had stronger affinity with psychiatry and psychoanalysis than with the hard core of academic psychology. And to a considerable extent, the more tenderminded in both anthropology and psychology have been the agents of contact. During the historical and descriptive phase of American anthropology, concern with theory was itself a mark of tendermindedness. In psychology, students of personality and social psychology, who comprised most of those who paid any attention to anthropological developments, were perforce bucking the current of methodological rigor in concerning themselves with these treacherous areas. In sum, when anthropologists looked to psychology, it was likely to be of the psychoanalytic variety, while the psychologists who were drawn into closest contact with anthropology tended themselves to be of the more clinical persuasion.[5]

[5] Margaret Mead's recent contribution to a textbook of psychoanalytic psychiatry at once discusses and itself exemplifies some of the affinities I have described. Cf. "Some Relationships Between Social Anthropology and Psychiatry," in *Dynamic Psychiatry*, F. Alexander and Helen Ross, eds., Chicago: University of Chicago Press, 1952, pp. 401–448. But Mead has also been concerned with underpinning her approach to personality-as-a-whole in culture-as-a-totality

Happily, this characterization, never entirely accurate, seems to be becoming outmoded by recent developments. Theory having become respectable in anthropology, tough-minded anthropologists are investigating problems that bring them into closer rapprochement with psychology. Psychodynamics having become respectable in psychology, toughminded psychologists are addressing themselves to motivational problems, and the gap between "academic" psychology and the more comprehensive but speculative theories of psychoanalysis is at least diminishing. As the attack on problems of motivation and personality development advances, moreover, the crucial relevance of the comparative data accessible through the anthropologist is becoming increasingly apparent.

Lest I be too quickly misunderstood, let me say at once that I have no wish to disparage the contribution of the tenderminded. But it is unhealthy for either the toughminded or the tenderminded to have a monopoly on particular problem areas or on the channels between two disciplines in the present stage of development of the human sciences. The present situation seems to me promising not because the rigor claimed by the experimental psychologist is prevailing, but because minds of both persuasions are more nearly meeting in the systematic attack on common problems.

While a genuine interpenetration of perspectives and collaborative marshalling of the joint resources of the two disciplines appears closer to realization in recent years, a less admirable sort of borrowing is perhaps still the rule at the plebeian level of the elementary textbook, where the most frequent if not the most significant occasions occur for the budding psychologist to form his impressions of related disciplines. The social psychologist who takes seriously the mission of his specialty to build a bridge between interpretations of individual behavior rooted in the organism, on the one hand, and phenomena at the social and cultural levels, on the other, must after all do more than merely borrow data and concepts. He has to re-examine the conceptual foundations of psychology to assure himself that they can anchor the individually-oriented end of this bridge effectively. To do this, he has to develop for himself a tentative conception of the structure of the bridge, including its anchoring at the

---

with appropriately rigorous field methods. See, for example, her chapter "Research on Primitive Children" in *Manual of Child Psychology,* L. Carmichael, ed., New York: Wiley, 1946, 667–706, which also contains a valuable discussion of anthropological contributions to the psychology of socialization.

socio-cultural end. More commonly, however, he has in fact tended to take for granted the existing conceptual structure oriented toward the biological individual, and without assuring himself that there is a bridge, to add on to it fragments and snatches casually drawn from the social sciences, particularly anthropology.

Solomon Asch, taking a critical view of the theoretical resources of social psychology, has recently made a similar point:

It has to be admitted that social psychology lives today in the shadow of great doctrines of man that were formulated long before it appeared and that it has borrowed its leading ideas from neighboring regions of scientific thought and from the social philosophies of the modern period. It is paradoxical but true that social psychology has thus far made the least contribution to the questions that are its special concern and that it has as yet not significantly affected the conceptions it has borrowed.[6]

Intellectual beachcombing of this sort is of course no substitute for serious attempts at synthesis in the social sciences. The presence in a psychological textbook of interesting anthropological case material or of genuflections to "culture" gives no guarantee that the contribution of anthropology has been assimilated. It may, on the other hand, contribute to the misimpression that a fully elaborated psychology would incorporate the entire domain of social science. As a sign of gropings toward a more comprehensive science of social man, it may temporarily be a healthy state of affairs, but it is a stage that needs to be superseded if we are to get ahead with the real collaborative task.

The Concept of Culture

Turning now to some of the specific contributions of anthropology to psychology, we can hardly start elsewhere than with the concept of culture, surely the most influential and perhaps the most important outcome of modern anthropology, particularly as it has developed in America. If the word "anthropology" were presented to a sample of psychologists in a word association test, I would venture, "culture" would probably be the most popular response, with "Indians" a runner-up. As a technical term, the word has filtered into the popular culture of educated people to an extent matched only by some of the language of psychoanalysis; the cautious explanation

[6] Solomon Asch, *Social Psychology*, copyright 1953 by Prentice-Hall, Inc., New York, p. viii.

that anthropologists mean by culture something other than the hoped-for product of a liberal education has become superfluous.

Anthropologists have the privilege of engaging in technical controversy about the definition of their most central concept[7]—controversy that leads to refinement and clarification when it is directed at an examination of common usage or of the logical requirements of the theoretical contexts in which the term is to be put to use, but which is apt to be sterile if it involves conflicting claims about the essence of a supposed real entity. But to a psychologist, there seems to be considerable agreement in the present use of the term. "Culture" in the abstract is a generalization from "cultures," while *a* culture is itself an abstract concept ascribed to some identified social group. The important components of current usage would seem to include (a) a conception of shared ways of behaving, predispositions to behavior, and (perhaps) products of behavior, and (b) the restriction that if something is a part of culture, it is learned—transmitted socially rather than biologically. Trouble enters primarily with the identification of what is shared. If culture is regarded as an abstraction from social behavior, however, the balance between patterned uniformities and individual variation becomes a matter for empirical determination rather than definitional controversy. In practice, the extent of uniformity or of patterning appears to be broad enough to make it convenient to talk about cultures, without dismissing the probable need for more differentiated concepts.

It is important to note that culture is a concept, not a theory. The term itself embodies no articulated propositions from which consequences can be drawn and put to test. It asserts nothing about reality. The psychologist who turns to anthropology after studying his lessons in the logic of science may therefore show some surprise when he encounters statements about the importance of the concept as an anthropological contribution.

It would be grossly unfair, of course, to view the concept of culture in splendid isolation. If it contains no assertions, its definition nevertheless calls attention to phenomena that had previously been neglected. With the *concept* of culture goes an *orientation*, if not a theory, that a very wide range of human phenomena is cultural in nature. The importance of the concept rests, then, on the host of

[7] Cf. A. L. Kroeber and C. Kluckhohn, *Culture: A Critical Review of Concepts and Definitions,* Cambridge: Peabody Museum Papers, Vol. XLVII, No. 1, 1952.

assertions in which it occurs, to the effect that phenomena x, y, z, etc. are cultural in origin, or are influenced in specific ways by culture. It is through facts of this latter order, together with the broader cultural orientation, that the concept of culture has been influential on psychology.

Some Parallels between the Concepts of Culture and Personality

In psychology, personality has much the same status as culture in anthropology, though it may lack the latter's eye-opening properties. As with culture, there is an entire literature of definitional controversy. To by-pass extensive discussion, we can take from Newcomb the modal definition that an individual's personality is the organization of his predispositions to behavior. Strictly speaking, *a* personality is to be distinguished from personality in general, just as *a* culture is from *culture*. Like culture, personality is an abstraction from behavior, mainly in social contexts. It is abstracted, however, from consistencies in the behavior of the single individual. Since personality has been linked with culture in one of the lines of development that has brought anthropology and psychology into close association, we can well afford a detour to examine jointly some of the controversies that have centered on these concepts. As it turns out, there is a striking parallelism in the methodological problems to which they have given rise. The same battles have been fought, apparently quite independently, in anthropology and psychology.

There is in the first place the question of how personality and culture are related to simpler or more microscopic concepts. The polarity of *elementarism* vs. *holism* that has run through the scientific controversies of the twentieth century has taken very similar form in the two disciplines, and in each the holistic emphasis on organization, hierarchy, and "patterning" has tended to gain the upper hand, after an initial period of predominant elementarism.

Anthropology of the diffusionist period concerned itself with the distribution of independent culture "traits" and "items," of which a particular culture was seen as an aggregate or congeries. While trait distributions continue to be studied, recent emphasis has fallen on the integrity of cultures. If the context in which particular traits occur determine their significance, to speak of the "same" trait in two different cultures is likely to require much qualification. A whole vocabulary has sprung up to deal with the hierarchical organization of cultural components, from "patterns" and "configurations" to "ethos" —even the more elementary terms having the flavor familiar to the psychologist from Gestalt doctrine. A similar trend is apparent in the psychology of personality. The "sum-total" conception of personality as a cluster of habits, S-R connections, traits or tendencies externally bound to one another has given way to emphasis on the personality "as a whole," with preoccupation with structure or organization.

In each case, however, the ascendancy of holism has raised a new set of problems that neither science has gone far toward solving. While it is the virtue of the holistic approach to recognize phenomena of structure and organization, either culture or personality "as a whole" is a more fitting subject for poetry than for science. Wholeness can be—and has often been—evoked, but it cannot be described in ways that can be checked independently without the use of more elementary terms. Yet if the sterility of crude elementarism is to be avoided, ways must be found of framing subsidiary concepts that are appropriate to dynamically organized phenomena, and techniques must be devised for constituting the appropriate data.

In this respect, the task of psychology may well be simpler than that of anthropology. The wholeness of the functioning organism, it has seemed reasonable to postulate, can readily be involved in complex responses, if, indeed, its involvement is not inevitable. In the development of projective tests, one attack on this problem, the psychologist has sought to obtain strategic samples of behavior from which the dispositional structure of the behaving individual can be inferred. Merest beginnings have been made in the process of grounding these tests more firmly in a rationale of psychological theory and standardizing the canons of inference from the data that they yield. But the persistence of knotty problems leaves reasonably clear the general direction from which a more adequate description of personality organization is to be sought.

The program of holism in cultural anthropology, on the other hand, must face the difficulty that since cultures are abstractions from the behavior of many people, they do not themselves "behave" in ways that can be sampled with a test, projective or otherwise. Analogous data have nevertheless been sought, from at least three directions. One may turn to the individuals who are vehicles of the culture, and borrowing techniques from the psychologist, investigate the organization of culture as it is internalized by its carriers. But the problem then remains of how to reconstitute "the" culture from these data. Or one may systematically analyze such cultural products as mythology and folklore to emerge with "themes" analogous to those that the clinical psychologist identifies in his projective material. Aside from the difficulty shared with the psychologist in arriving at objective canons of inference, the anthropologist faces here a major problem of attaining a definable and adequate sample of the culture for his analysis. Or, finally, one may look for the covariation, relative persistence, and modifiability of cultural components under conditions of culture contact, change, and dissolution. The latter method, attractive since it identifies structure in what is functionally coherent, is applicable only when it is possible to capture adequate data on culture change. Each of these approaches has its instructive parallel in the psychology of personality. As either discipline approaches more closely the solution of its problems of structural description, the other may find leads from it toward the solution of its own problems.

A second realm of controversy has to do with the sort of theoretical context in which culture or personality concepts are to be placed —or, to put it differently, with the explanation of cultures or personalities once they have been described. The polarity, as it became defined in anthropology, is that of *historicism* vs. *functionalism*.

In both disciplines, there has been affinity between historical interpretation and the preference for working with simple and unchanging conceptual elements. The association is a natural one, since if personality or culture is regarded as a chance assembly of elements that preserve their identity and are not significantly influenced by their interaction with one another, the presence of a culture trait or a habit or S-R bond can only be accounted for historically. The presumed rigid identity of the element draws the theorist's attention to the historical line of continuity in its past vicissitudes, and underplays the lines of relationship to features of the contemporary situation.

If the culture trait is independent of the organized cultural matrix in which it is embedded, one can only refer its presence to persistence from an earlier state of the culture, or to accession from processes of invention or diffusion. Conversely, the independent trait as a starting point gives little anchorage for investigating the functioning of socio-cultural systems or studying the relation of cultural components to the presently existing needs of the members of a society. So functionalism of the varieties associated with both Radcliffe-

Brown and Malinowski has developed within a descriptive orientation that weights heavily the phenomena of organization and structure.

In psychology, too, elementaristic description has involved predominantly historical explanation: witness the historical phrasing of classical associationism and of its more modern variants in conditioning and reinforcement theory. But the existence of strong functional and structural orientations in psychoanalysis, which has been notably historical in its preoccupation with origins, suggests that the antihistorical bent of functionalism in anthropology may have been accidental rather than the outcome of intrinsic features of a functional approach. Modern psychology, historical or otherwise, is in fact overwhelmingly functional. The dominant strain is oriented toward a model of the organism as a self-regulating system and falls naturally into the use of terms like homeostasis, equilibrium, and adjustment; while the marginal influence of Gestalt theory leads to parallel emphasis on the field determination of phenomenal properties or behavioral tendencies. Rather than putting history and function in opposition, recent psychological discussion has centered on the more clearly articulated dichotomy of *historical* vs. *contemporaneous* explanation, posed most forcefully in the writings of Kurt Lewin.[8]

And the history of even this dichotomy, if I read it rightly, has indicated that once certain tenacious misunderstandings are cleared aside, there is little cause for battle. If psychology were equipped with an adequate set of laws and an adequate description of the situation faced by an individual, as most psychologists would be ready to agree in principle, his behavior could be predicted either from a complete record of his psychological history or from a full analysis of his immediate predispositions—the present state of his personality as a system. In this hypothetical case, which for practical purposes is trivial because it is non-existent, knowledge of the individual's history would be translatable into knowledge about his contemporaneous personality structure. Whether or not one would then elect to work in historical terms would be a matter of arbitrary preference.

The fact of the matter, however, is that psychologists do not and will not in the foreseeable future possess complete data on either the history or the present personality of an individual, to say nothing of

[8] Cf. his *Dynamic Theory of Personality*, New York: McGraw-Hill, 1935.

adequate laws or procedures for situational analysis. Lacking the equivalent of X-ray techniques to diagnose present personality, the psychologist perforce has recourse to personal history in his recon-struction of the individual as he is at present. Correspondingly, re-constructions of personal history must usually take as their starting point the individual's present dispositions, including most notably his memories.

There is thus no inherent conflict between historical and ahistori-cal approaches to the study of personality. They are complementary, not antagonistic. Once the air is cleared of the murk of partisanship, investigators can get about their business, some approaching the problem from a focus on antecedent—consequent relations in indi-vidual history, others seeking to analyze functional interdependen-cies within the immediate present, and still others working toward a closer articulation of the two approaches. The lessons of this contro-versy, which has perhaps been worked through in greater detail in psychology than in anthropology, should be partially transferable to research on culture. Indeed, Murdock has already demonstrated that analysis of functional dependence among elements of social structure furnishes a powerful tool for historical reconstruction.[9]

The third methodological difficulty running alike through discus-sion of personality and of culture has also been phrased as a polarity, that between *idiographic* and *nomothetic* approaches. The terms were given currency in psychology by G. W. Allport,[10] who derived them from Windelband. The contrast is that between fidelity to the unique occurrence and the search for general laws that encompass reality by successive approximation without pretending to capture the unique. An idiographic psychology, as proclaimed by Allport, would seek to describe the personality of each individual in terms of its own unique organization. Its quest for laws would be directed at the discovery of principles of development and organization to account for individual uniqueness. The corresponding position in anthropology reiterates the uniqueness of cultures, and shows an aversion to systematic comparative analysis as somehow involving the violation of a culture's integrity. There is, indeed, a strong cur-rent of temperamental preference running through the statements of faith from both sides of this controversy: the more tenderminded and intuitive aligned on one side with a nurturant, appreciative atti-

[9] Cf. his *Social Structure*, New York: Macmillan, 1949.
[10] *Personality*, New York: Holt, 1937.

tude toward their data, as against the toughminded and conceptual who have no hesitation in rending apart the presenting phenomena in order to abstract regularities that can be fitted into a causal though approximate interpretation of nature.[11]

In fairness to Allport, it should be noted that he regards the idiographic and nomothetic approaches as complementary. Those who embrace the extreme idiographic position, which I believe is more deeply entrenched in anthropology than in psychology, are vulnerable to a telling argument. If they maintain that cultures or personalities are each unique and incommeasurable, one has only to counter with instances of success in the demonstration of lawful relationships in cross-cultural studies or in experimental psychodynamics to make the position untenable. Meager as our substantive findings are today, they more than suffice to establish this point.

The dichotomy collapses, it seems to me, when it is recognized that all concrete phenomena, whether sticks or stones or behaving individuals in their interrelations, are unique. One may quite legitimately appreciate their uniqueness and even haltingly aspire to describe or recreate it, though researchers of this bent seem to be drawn toward poetry and metaphor. But one may also legitimately seek to abstract from phenomena common analytical elements that can then be lawfully related. The lesson of the history of science is that this detour away from the compelling force of concrete reality greatly increases our powers of understanding it and coping with it in the longer run. Quite apart from the general question of the definition of science, however, it is clear that the psychologists, sociologists, and anthropologists who are groping toward a more comprehensive behavioral science are committed to pushing the latter sort of program as far as it can be extended.

But the idiographic position deserves serious consideration in the present connection to the extent that its arguments merge with the admonitions of holism. If context matters, as it presumably does, serious difficulty lies in the way of comparative, nomothetic, research

[11] Like most typologies, the present one—introduced by William James— does some violence to the facts. Among ethnologists what is certainly an idiographic approach is often pursued with the most ascetic toughmindedness. Where fact reigns supreme, any attempt at theory may appear tenderminded. The present distinction holds rather among different persuasions of the theoretically inclined, dividing those who look for an abstract formulation of particular cultures from those who are interested more in cross-cultural regularities. In psychology, the concrete, factual sort of toughmindedness is little represented. Good or bad, psychologists seem all to be theorists.

that neglects contextual effects in the definition of its variables. In both personality research and research on cultures, and especially in the research area that deals with the interrelations of culture and personality, one of the most urgent methodological tasks is the identification of variables that are truly homologous in different contextual settings. Such dimensions of personality and variables of culture patterns, as they are progressively refined, should bring increasing areas of untamed natural uniqueness under scientific cultivation. Good beginnings are already being made in this direction. While proponents of the idiographic position will not contribute directly to this essential task, they remain valuable critics who may be relied on to remind us of how little and how tenuously the jungle has been tamed, and, equally important, to give us preliminary accounts of some of the strange forms that lurk there.[12]

Lastly, the very task of defining culture and personality has been beset in each case with uncertainty about the logical status of the terms. Is culture, and is personality, something real in nature whose lineaments we have only to discern truly? Or are these merely terms of convenience, that we use as we like with no implied counterpart in the structure of reality? The foregoing discussion has assumed the now prevailing view which differs from both of these respective realist and nominalist positions. Personality and culture as terms in scientific discourse are both constructs, figments of the scientific imagination. But if they are to play a useful part in the scientific

---

[12] An additional methodological parallel might be drawn in regard to the role of typological formulations about personality and culture. Typologies have appeal as a compromise between holistic and idiographic concern with fidelity to the particular case in its structured uniqueness, and the attempt to attain a degree of generality. A half-way station on the road to science that has usually had more heuristic convenience than empirical validity, the typological approach represented in psychology by Kretschmer (constitutional types), Jung (introversion-extraversion), and Spranger (economic, theoretical, political, social, and religious value-types) has given way to the use of interrelated variables.

Unlike sociologists, anthropologists have not made very much use of typological concepts. Examples, perhaps, are Redfield's adaptation of Toennies' concepts of *Gemeinschaft* and *Gesellschaft* in his treatment of folk vs. urban societies, and Ruth Benedict's distinction between Apollonian and Dionysian cultures. Typologies based on kinship terminology are a different case, since the social and biological distinctions around which kinship nomenclature and practice are elaborated are discontinuous. The limited range of possibilities therefore makes the data fall naturally into clear types. But if anthropologists have not been given to typologizing, neither have they done much with explicitly interrelated variables. Until culture theory is more highly developed, the half-way station of typologies may perhaps continue to have its usefulness.

enterprise, the scientist must be highly restricted in his freedom to have his will with them. He must frame and elaborate them with due regard to the requirements of the theoretical context in which they are to be put to use, and this context of theory, in turn, must stand the test of empirical verification. So, in an indirect way, constructs are grounded on reality to the extent that they are embedded in a maturing theory in constant interplay with data. The crucial fit is between theoretical structure and the data that confirm or refute it, not between the several constructs and presumptive entities taken as things-in-themselves.

This perspective, a conceptualist one if you like, is becoming increasingly explicit and prevalent in psychology and the social sciences. It has a number of important implications for theory-building and research. As constructs, personality and culture are abstractions. To recognize this is to undercut the greater part of the nomothetic-idiographic controversy. It is things-in-themselves whose uniqueness is defended by the idiographically inclined; the worth of a construct is measured, as we have observed, solely by the extent to which it takes its place in significant propositions that can later be verified. Whether it corresponds or not to unique reality has nothing to do with the matter. But note again that the value of the construct cannot be ascertained in isolation. Only as part of a theory can it be assessed. Much of the "theoretical" discussion of both personality and culture has remained at the level of elaborating concepts, without making explicit the propositions from which they can alone derive significance.

More differentiated concepts at the level of motive, value, custom, or culture pattern may well turn out to be more fruitful than culture in the generation of hypotheses and laws. The global concepts of culture and personality may then end up after all as terms of convenience like mechanics and thermodynamics to identify major research areas, rather than as serious constructs of scientific theory.

In the present state of anthropology and personality theory, however, explicit recognition of the status of personality and culture as constructs is most salutary for the clarification that it brings into discussion in the fashionable culture-and-personality area. Rather than being concerned with the relations between two mythical interacting entities, one is led to look afresh at the different angle of abstraction by which each construct is derived from the data of social behavior. The substantively empirical findings can be assimilated more effectively into the theoretical frameworks of the respective disciplines when one is equipped with a translation formula. Some of the relations between the two concepts, as we will note later, turn out to follow from logical necessity.

The Influence of the Concept of Culture on Trends in Psychology

Our excursion to look at some problematic features of thinking about culture and about personality has laid before us some of the methodological complexities involved in a sophisticated treatment of culture. The gross impact of the concept of culture on psychology, however, has occurred at a less sophisticated level. It is as a general orientation to the determinants of human behavior rather than as a carefully elaborated concept or a developed theory that the cultural approach has been most influential.

Culture and the Influence of the Environment. Recognition of the extent to which the activities of the members of a society are culturally patterned and of the extraordinary range of variation in modal patterns as one moves from society to society led to a veritable revolution in psychological thinking that is still working itself out today. Darwinian biology had brought the importance of environment to attention quite early in the development of modern psychology. But so long as all psychological investigation was carried out in the relatively standard environmental range of Western civilization, it was easy to mistake the consistent results of exposure to this setting for features of a "generalized human mind" or general principles of behavior—a natural fallacy that persistently recurs, as when a Kinsey entitles the report on his researches in American sexual behavior "Sexual Behavior in the *Human* Male." Anthropology raised serious questions as to whether the entire structure of the psychologist's supposedly universal-human doctrine might not be in important respects "culture-bound." The latter term, in fact, compresses to a word the major burden of anthropological criticism.

So well entrenched was the narrowly based attempt at a generalized human psychology in the early twentieth century, new and fragmentary as were its facts, that one early reaction to the beginning influx of accounts of behavior in non-literate societies was to read in them evidence for a child-like primitive mentality. If "natives" did not think like Western adults, obviously they were to be equated to Western children. Western man still furnished the unquestioned standard, and the result did not disturb the prevalent complacency that placed the modern Englishman or German (it varied with the nationality of the author) at the peak of the evolutionary scale. As the last remark suggests, legacies of the initial uncritical response to the Darwinian revolution made this blind alley attractive. The anthropologists of the day were assuming that societies could be placed along a unilinear scale of cultural evolution, while psychologists like G. Stanley Hall had made popular the view that the development of the child recapitulates in microcosm the evolution of the race. Although the doctrine of the primitive mind was promulgated principally by non-psychologists like Levy-Bruhl, it was in keeping with the culture-bound psychology of the day.

Subsequent criticism, of course, has questioned both the description of non-literate thinking (it may often be rational) and that of the thought processes of the Western adult (which seldom follow the rules of logic), as well as discrediting the framework of evolutionary assumptions that made it plausible to equate the child and the primitive. Revulsion against this tangle of error, unfortunately, has diverted attention from genuine problems that the doctrine of primitive mentality approached but wrongly phrased. As Werner has suggested in an unfortunately neglected work[13] that partakes to some extent in the earlier fallacies, there may well be intrinsic sequences in the mental development of the child, involving progressive differentiation, conceptualization, and articulated integration as the child is exposed to environments of the sort represented in the extreme in Western urban society. If there are indeed psychological grounds for expecting a developmental sequence, the ordering though not the pacing and terminal level of which is relatively independent of culture, critics of primitive mentality may have been too hasty in rejecting concern with developmental levels. In point of fact, there are scarcely any existing data on the development of cognitive processes in differing cultural contexts. Serious comparative research using modern psychological techniques is badly needed to test out the possibility of a truly general though culturally relevant psychology of cognitive development.

It was in the attack on instinct doctrines that the cultural approach made its most important contribution to a revised apprecia-

---

[13] Heinz Werner, *Comparative Psychology of Mental Development,* rev. ed., Chicago: Follett Publishing Co., 1948.

tion of environmental influence, both in academic psychology and, more recently, in psychoanalytic theory. The lists of purportedly universal-human instincts elaborated by James, McDougall, Thorndike, and others were not long in crumbling when confronted with the surprising differences in activities and motives reported from the majority of societies that fall outside the purview of Western history. Once the structure was breeched, it became apparent that even highly "dependable" motives like parental love and gregariousness could as well be attributed to universal features of the human environment (first learnings always occurring in the bosom of the family) as to universal constants of biology.

Self-isolated from these scientific controversies and combining sectarian dogmatism with pragmatic disregard for the niceties of scientific confirmation, Freudian psychoanalysis was still resisting the impact of anthropology after academic psychology had capitulated in principle. It clung tenaciously to a biological libido that unfolded in ways biologically foreordained, including the complicated configuration of the Oedipus complex. While Freud and, later, Roheim drew on ethnology to support a suggestive interpretation of culture in orthodox psychoanalytic terms, one that erred seriously as it pretended to historical reconstruction or to unilateral derivation of cultural from psychological facts, the more advanced thinking in all psychoanalytic camps is at present coming to realize the importance of disentangling what is general in psychoanalytic doctrine from what was actually peculiar to *fin-de-siècle* Vienna or to the broader range of Western culture. The line of criticism initiated by Malinowski, for example, has given the Oedipus complex a less culture-bound interpretation: the patterned residue of the person's early affectional relations with the people who nurtured, disciplined, and socialized him in the family. As the family is universal, so is the Oedipus complex in this broadened sense, which sets the stage, however, for the search for differences in its patterned content corresponding to the wide variety of family relationships that are somewhere culturally prescribed. Recent research on "culture and personality" is only beginning to explore systematic relationships in this area.

We can pass over the extreme environmentalist phase into which the behavioristic psychology of the 1920's swung in reaction against the errors of instinct theory, observing only that psychologists seem to agree today in regarding personality and behavior as complex resultants of a chain of organism-environment interactions, of which genetic and environmental determinants are both inescapably necessary. Anthropological data and insights have also highlighted the importance of environment in a somewhat different sense. While the investigator may attempt to specify a person's environment in objective terms independent of the special perspective of the behaving individual, human beings after early infancy have commerce with such a "real" world only through mediating processes of selection, elaboration, and interpretation made possible by the highly developed symbolic equipment that is at once the badge of humanity and the core of the cultural heritage. The process of mediation is so complex and so little understood that for most purposes a more intelligible account of human behavior can be gained by referring it to the "psychological environment," the "private world" in which the individual is oriented, than by attempting to start with the objective environment of the physical scientist and biologist.

The concept of the psychological environment grew current in psychology through the writings of Koffka and Lewin. Hallowell[14] has reminded us that the data of modern anthropology demand a similar conception of reality as it is defined for the individual by the premises of the culture in which he participates. For "private worlds" are not so private after all; much of their distinctive structure is shared among all people who grow up to speak the same language, believe in the same gods, and, through inheriting a common culture, come to see as through lenses ground to a common formula. These are facts with which the psychologist has to come to terms, difficult as he may find it to discover the transformation formulae relating such inferred subjective structures to "stimuli" within the objective frame of reference. As out-and-out objectivism and interpretation in terms of the psychological environment—sometimes rather unhappily termed the "phenomenological frame of reference"[15]—compete in contemporary psychology, the weight of cultural anthropology tends to fall on the side of the latter. While the phenomenological

[14] Cf. A. Irving Hallowell, "Psychological Leads for Ethnological Field Workers," in D. G. Haring, ed., *Personal Character and Cultural Milieu*, Syracuse, N.Y.: Syracuse University Press, 1949, pp. 290–348, especially pp. 308 ff.; also his article "The Self and Its Behavioral Environment," in *Psychoanalysis and the Social Sciences*, Vol. 4, G. Roheim, ed., New York: International Universities Press, 1953.

[15] Cf. my article, "The Phenomenological Approach in Personality Theory: Some Critical Comments," *Journal of Abnormal and Social Psychology*, 1950, 45, 516–522.

psychologist may too readily take the psychological environment as given, the objectivist risks sterility and irrelevance if he evades accepting it as a problem.

Looking back at the revolution touched off in psychology by the cultural approach, the surprising thing is the still limited extent to which it has come to fruition. Psychologists continue to aspire to a general science of behavior—no one responded to the cultural revolution by happy resignation to the explicit study of, say, the special psychology of Americans. Yet overwhelmingly their experimental studies are confined to subjects from modern Western culture or from a trifling number of sub-human species. Of late there has been a resurgence of neo-comparative psychologists who have decried the founding of "general" principles of behavior on the white rat and the chimpanzee. All the more should it be essential, not only for social psychology and personality study but for the development of general psychology as well, that fundamental psychological experimentation be extended comparatively to subjects from a wide variety of cultures. Though dramatic, the impact of the cultural approach on psychology has occurred principally in regard to instinct theory, the interpretation of abilities,[16] and personality study. Surely such important areas as perception and the thought processes, learning, and "group dynamics" need re-examination from a comparative, cross-cultural perspective. The existing anthropological literature suggests that substantial revision of current doctrine might result.

Cultural Relativism. To recognize that each culture defines its own version of reality, its own standards and ethical system, and imposes its own distinctive patterns as "normal" is to take the position of cultural relativism, once modern civilization is accepted as one culture among others rather than taken as in some sense peculiarly human. In terms of lip-service, cultural relativism is almost as widely embraced in psychology as in anthropology. Already avoidant of value judgments, psychologists have found in the relativistic point of view additional armor for safeguarding scientific objectivity and ethical neutrality.

A decade or so ago it would have been easier than it is today to assess the significance of cultural relativism for psychology. One would then have pointed with some assurance to a resulting revision of concepts of psychological normality then in progress. The revision,

[16] Not treated here because of its lesser relevance to existing systematic theory.

which we now see left some important problems unresolved, was nevertheless a real and needed one: it turned out that two important meanings of this elusive term were implicitly cultural and had better be made explicitly so. On the one hand are the statistically frequent behavior patterns of a group; on the other, the group's ideals of approved behavior. It certainly cleared the air to recognize that both conceptions of normality—and only confusion results unless they are kept distinct—pose as alternatives either a relativism that makes the cultural reference explicit, or a dogmatic absolutism that is in fact culture-bound. In psychopathology, it became pertinent to re-examine the culturally-given standards of our own society before applying them ethnocentrically to the analysis of bizarre behavior patterns of other cultures.

But psychologists could not be entirely satisfied with the starkly relativist solution that seemed at first to cut through so many problems. In the extreme version, cultural relativism involves an idiographic approach to cultures that excludes the possibility not only of trans-cultural norms and standards, but of general laws and principles arising from cross-cultural comparison. Mixed in with the admittedly cultural meanings of normality as it occurred in psychological parlance were gropings toward more intrinsically psychological criteria of adequate functioning. These gropings have continued unabated, and recently anthropologists as well as psychologists and psychiatrists have given increasing attention to ways of transcending the culturally relative, without denying the importance of the relativistic critique of traditional absolutes. The flavor of the current search for transcultural standards of psychological functioning, inconclusive as it has thus far been, can best be evoked by listing some of the terms under the aegis of which the quest is in progress: maturity, autonomy, competence, productive orientation, mental health. "Adjustment," unless very specially defined, should perhaps be conspicuously absent from this list, since one has only to ask what the person is adjusted *to* for the cultural reference to become obvious and the old alternatives of relativism vs. culture-bound absolutism to be reintroduced.

Volumes have been written on each of these concepts and others like them. Here I can only indicate the direction along which I think progress can be made toward a solution. First, under whatever term psychological adequacy is assessed it seems likely that no single criterion can be sufficient for its evaluation. A set of imperfectly correlated criteria employed jointly seems essential. Value judgment, moreover, is unavoidable in their selection. Under the impact of increasing communication among different value perspectives and in the light of growing psychological and cultural knowledge, there is nevertheless hope for an emerging informed consensus about the marks of psychological adequacy. Say that such criteria turn out to include indices of attributes like creativity, flexibility, self-acceptance, effective intelligence, etc. Then, secondly, characteristic average values of each of these variables can be found for persons growing up in distinctive cultural milieus.

More important, in most actual cultures, an individual can be expected to attain optimal values on certain criteria only at the cost of less than optimal realization of others. In a culture that is characterized by conflicting values as ours is said to be, for instance, adaptation may be attainable only at the cost of personal integration, and vice versa. One can thus look toward the assessment of cultures in regard to the extent to which they permit the joint realization of a plurality of criteria, as well as the assessment of individuals in regard to their standing on the criterion variables.

Since the approach rests fundamentally on evaluative criteria established by consensus, competing patterns of analysis are appropriate as long as value frameworks diverge. But we must abide by this pluralistic state of affairs if we are to avoid traditional or theological absolutism. It differs from the older extreme relativism, however, in its intended movement toward standards of evaluation that are progressively free from restriction to single cultures. Transcultural standards can be approached only with difficulty and through slow successive approximation; there are no lurking rabbits ready to be pulled from the hat by an ingenious maneuver. Whether we like it or not, social science is no *deus ex machina* to replace the God of our fathers in generating full-blown a new set of absolutes. But neither is it irrelevant to the clarification of values.[17]

Culture "and" Personality. The "culture and personality" movement, which has progressively gained momentum since its launching in the thirties, merits special consideration as the interstitial development that has brought anthropology and psychology into closest contact. The very existence of a movement so designated points up

[17] A more extensive essay in this vein may be found in my article, "Optima of Mental Health: A General Frame of Reference," *Psychiatry*, 1950, 13, 503–510.

the excessively narrow framework within which investigations of culture and of personality had been respectively pursued in the two disciplines. Except as a disembodied abstraction, culture can hardly be reckoned with apart from persons; from the other side of the fence, personality cannot be viewed in perspective apart from the cultural tradition that furnishes so much of its form and content. The movement was thus an attempt to join together what should never have been held asunder in the first place. While the literature identified with it has tended toward preoccupation with a narrower selection of problems, the broadest interpretation of its objectives would be almost as inclusive as the scope of the present volume.[18]

By and large the initiative was taken by anthropologists with collaboration from psychoanalysts (notably Abram Kardiner and Erich Fromm), and rather less participation by psychologists. But the problems dealt with are as relevant to psychology as they are to anthropology. While the anthropologist following his traditional interests might look for psychological clues to the processes of cultural continuity and change and to the cohesiveness and range of variability of culture patterns—the dynamics and microstructure of culture—the psychologist's principal concern centers on how cultural factors enter into the determination of personality structure and into the definition of situations in which the developed personality functions. The former of the psychologist's potential interests, it should be added, has almost eclipsed the latter, which in principle should be equally pertinent.

There is no space here to review the literature on cultural factors in personality formation, which has been rich both in suggestive but

[18] Bibliographies of this literature can be found in the contributions by A. Irving Hallowell and Margaret Mead to the Wenner-Gren symposium of 1952. See respectively "Culture, Personality, and Society," and "National Character," in *Anthropology Today*, by A. L. Kroeber and others, Chicago: University of Chicago Press, 1953, pp. 597–620, 642–667. Notable collections of papers are Douglas G. Haring, ed., *Personal Character and Cultural Milieu*, Syracuse, N.Y.: Syracuse University Press, 1949, and Clyde Kluckhohn, Henry A. Murray, and David Schneider, eds., *Personality in Nature, Society, and Culture*, rev. ed., New York: Knopf, 1952, the former also containing an extensive bibliography. A symposium discussion of the field may be found in S. Stansfeld Sargent and Marian W. Smith, eds., *Culture and Personality: Proceedings of an Interdisciplinary Conference Held Under the Auspices of the Viking Fund, November 7 and 8, 1947*, New York: Viking Fund, 1949. Otto Klineberg's *Tensions Affecting International Understanding*, New York: Social Science Research Council, Bulletin 62, 1950, surveys the literature on national character, with particular attention to methodological problems.

fragmentary data and in attempts at theoretical synthesis. I will rather address myself briefly first to an intentionally sweeping characterization of the research situation in relation to present needs, and secondly, to an attempt to clarify some of the conceptual problems that bother a psychologist as he seeks in this area a bridge between his home territory and the social sciences.

It is probably fair to say that anthropological research in culture and personality to date has provided us with highly provocative case material and leads, without, for the most part, the systematic evidence essential for confirmation. There is ample evidence that there are bear in these woods, enough to justify major concerted efforts to track them down. But the tracking operation has been only begun. The reason for this state of affairs is not hard to find: it lies not so much in the deficiencies of research efforts thus far as in difficulties inherent in cross-cultural methodology.

Consider for a moment the case approach in personality study. A skilled clinician, working intensively with a single individual, can build up a highly plausible construction of the causal interplay that has given rise to his present personality. Such a construction can carry great conviction, particularly if there is internal consistency among converging lines of evidence. But it stands up rather poorly against obdurate criticism that insists on the exclusion of alternative and equally plausible interpretations. And, more important, the single case does not yield general principles (here the idiographic-nomothetic controversy comes in). These must be drawn from the study of systematic covariation in series of cases, ideally under experimentally controlled situations. It is from cumulative experience with series of cases that the major psychodynamic insights have come, and it is from more controlled studies approaching the experimental ideal that they are being validated and refined to take their place in developed theory.

But a field report on "culture and personality" in a single society contributes the equivalent of a single case, even if studies of a number of individuals form part of the research. With the best techniques, it can establish beyond question that certain cultural practices and norms—in regard to child rearing, life crises, or whatever—are associated in this instance with certain distributions or modal tendencies in the personality patterns found in the group. The causal interpretation of such obtained associations, however, cannot be substantiated by internal evidence. For this, systematic series of cases—of field studies in different cultures—are essential. Yet the cases accumulate very slowly, even when field work is conducted much more hastily and casually than the complexities of the data would warrant. Small wonder, then, that culture-and-personality theory remains controversial.

Most of the field work with which we are concerned has of course been undertaken with a different point of view toward the task. Working hypotheses and principles have been borrowed from psychodynamics and learning theory and applied to the interpretation of cultural data. There is nothing exceptionable in this procedure, which as we know has redounded to the benefit of personality theory and insight into cultural processes alike. But so long as the doctrines being applied are controversial, their application to the explanation of personality development in a single culture must remain even more so; not only are the principles themselves still in doubt, but there is added uncertainty as to the correctness of their application to the facts at hand. Good fit in the particular case contributes, to be sure, toward the validation of the interpretative principles that are employed, but it does not go very far toward establishing them beyond cavil. And cavil there has been.

The way out of the hardly defensible bootstraps dilemma of seeking simultaneously to validate interpretative principles and interpretations against each other is to move from the single case to the case series. With a statistically adequate sample of cultures, hypotheses drawn from psychological theory can be checked against correlations obtained between cultural variables and variables of modal personality. Since the cross-cultural range of variation in conditions of socialization is much more extensive than that in Western society in which theories of personality development originated, extension as well as revision of existing theory can be anticipated.

Impressionistic comparison of the accumulating body of cultural cases, while superior to reliance on internal consistency within single studies, is still inadequate to the task. For the most part the studies were never conducted with an eye to comparability, and far too many degrees of freedom are left to the insightful scholar who seeks to distill their cumulative message. A half-way station is represented by the attempt to extract quantitative relationships from existing materials by applying objective rating methods or typologies, and statistical tests of significant association. The resources of the Human

Relations Area Files,[19] which organize in accessible form the major descriptive literature on some 200 cultures, make it possible to go some distance in applying to problems of socialization and personality development this approach, the power of which was demonstrated in another area by Murdock.[20] The recent work by John W. M. Whiting and Irvin Child,[21] who studied the relation of infant training in spheres suggested by psychoanalytic and learning theory to the projective reflection of adult anxieties in magical explanations of disease, carries us considerably ahead of the prevailing impressionism. No doubt they have far from exhausted the potentialities of the Human Relations Area Files for culture and personality research.

The stage nevertheless seems set for a frontal attack through original field work in an adequate sample of cultures to collect systematically the directly relevant data for testing the relationship between different patterned modes of child rearing and selected personality variables. Only by ingenious indirection were Whiting and Child able to explore such hypotheses in a preliminary way in the existing literature. Their work gives promise of even more substantial advance from research planned so as to yield indices that are based on the most direct evidence obtainable through comparable methods. Technical difficulties remain to be worked through in regard to both the isolation of transcultural variables and the construction of practicable indices, but progress is being made on both scores.

When it comes to causal interpretation, however, even the most ideal design we can conceive for cross-cultural research on personality formation runs into the circular chicken-and-egg relationship that we must assume to hold between cultural and personality variables. One of the current insights that is hardly to be controverted has it that in stable societies parents raise their children in ways that lead them to become in turn the kind of parents who naturally want to raise *their* children likewise. Mere correlations between rearing practices and adult personality do not disentangle a primary causal direction within this presumptively circular system. The matter is

[19] Cf. G. P. Murdock, C. S. Ford, et al., *Outline of Cultural Materials,* 3rd rev. ed., New Haven, Conn.: Human Relations Area Files, Inc., 1950.

[20] G. P. Murdock, *op. cit.*

[21] J. W. M. Whiting and Irvin Child, *Child Training and Personality,* New Haven: Yale University Press, 1953.

further complicated by the influence on personality of cultural pressures encountered throughout the life cycle. Important as we have grounds for believing infancy to be, we can be sure that socialization does not stop there. These later influences, if themselves correlated with parental practices, could underlie outcomes mistakenly attributed to infantile experience. Even systematic cross-cultural comparison can therefore offer no royal road to uncomplicated truth. Sometimes we may have sound theoretical grounds for assigning a single direction of causation. And intensive studies of changing cultures may loosen the strands of the causal nexus sufficiently for them to be partially disentangled. For the most part, however, the causal interpretation of the associations turned up by research is inherently ambiguous.

Actual research on culture and personality has focused on process relationships among cultural and personality variables, not on culture "and" personality as presumptively separable entities. Hence the purist's need to put the "and" in quotation marks. Yet some of the theoretical discussion of culture and personality has been seemingly beguiled by the conventional label into neglecting the fact that both abstractions rest ultimately on the same data of social behavior. Once definitions of culture and personality such as those offered earlier are accepted and it is stipulated that phenomena exist to which the definitions can be appropriately applied, some of the matters about which there has been no little controversy turn out to follow tautologically.

So it has been with conceptions of modal personality, basic personality, and their ilk. If in fact the members of a definable social group share a set of traditional behavior patterns that warrant designation as a culture or sub-culture, by the same token these patterns (having no existence apart from behaving persons) cannot fail to be integrated into the personalities of the members. Cultural universals, to employ Linton's term, must turn up in the personality dispositions of virtually all the group members, else they would not be universals; these are ingredients of basic or modal personality. Cultural specialties, such as those that make it relevant to distinguish particular positions or statuses in the society, may be referred accordingly to dispositions shared by the occupants of the respective statuses; Linton has given us the term "status personality" to call attention to this culturally relevant aspect of personality patterning.[22] The "existence"

[22] Ralph Linton, *Cultural Background of Personality.*

of modal personality and status personality is no more controversial than the existence of cultures and statuses. At root these are merely alternatives of organizing the same data.

If this were all there were to it, however, the concepts would never have aroused the degree of interest that they rightly attract. The interesting features of modal or basic personality, which are far from tautological, are secondary repercussions arising through intrinsic psychological processes in response to the impact and internalization of culture patterns in personality development. Tendencies for parents to respond harshly to their children's aggressive behavior, for instance, if common to the members of a society, are to be referred equally to the culture and to the modal personality of the parents. But the result in the developing child is not a foregone conclusion: present knowledge suggests that under specifiable conditions outcomes as different as rigid politeness or touchy latent hostility may follow. These consequences in turn may lead to cultural elaborations that seem superficially remote from the cultural starting point, yet are dynamically linked with it: a preoccupation with sorcery and witchcraft, for example, arising as an avenue for the discharge of pent-up hostilities. The situation is further complicated by the likelihood that these secondary cultural repercussions, while reinforced by the psychodynamic consequences of other cultural practices, may themselves be learned directly by the participants in the culture. And so we are thrown back into the intricate causal interplay that we noted as complicating the interpretation of cross-cultural research. It is nevertheless well to remember that while the elaboration of modal personality and its ramified implications for the coherence of cultural organization can only be discovered through empirical research, the logical status of the concept is such that whenever sufficient regularities exist to identify a culture, we may be sure it is applicable.

The tendency to reify concepts, to give them "misplaced concreteness," which is a persistent source of confusion in this area, has potentially unfortunate results on personality theory as it digests the implications of culture-and-personality research. It is legitimate and helpful to classify the determinants of personality according to their source in facts of biology, culture, or social role, and to abstract a conception of modal or status personality. But it is misplaced concreteness to assume uncritically that these analytic distinctions are relevant to the functioning of the integral person. Yet one often gets the impression from current writing that personality can somehow be partitioned into concentric segments identified by their source of determination—a biological core, a region of modal personality, and so on. This sort of construction, which is implicit in the frequently encountered definition of personality as the source of residual variation once culturally determined regularities have been extracted, makes no sense psychologically. Psychological accounts of personality functioning need to draw their own relevant distinctions, mindful, of course, that to understand the personal significance of any item of behavior one must know the cultural norms toward which it is oriented.

Some Directions of Convergence

I have already hazarded a number of suggestions about ways in which the collaboration of anthropology and psychology might become more fruitful. There is no need to rehearse them here. Some concluding remarks are nevertheless in order to make some additional observations and point up directions of convergence that are implicit in my treatment.

Broadening the Base of Collaboration: Language and Psychology. Research on culture and personality has tended to concentrate, we observed, on a restricted range of problems. A broadened base of collaboration between anthropology and psychology would bring additional areas into joint focus. So far as psychology is concerned, there is no area in which the gain from collaboration should be greater than that of language research. For the psychologist it is still almost virgin territory.

This is the case in spite of universal agreement among psychologists about the distinctive and fundamental importance of language in human behavior. Aside from an impressive body of research on language development in European and American children—which did not have to await the construction of adequate theory—psychological treatment of language and symbolic behavior has been speculative and relatively barren. Themselves immersed in language, psychologists have seemed unable to discover leverage points at which research could clarify the nature of talking and listening and verbal thinking as psychological processes. And their theoretical speculations and halting attempts to experiment have proceeded in unfortunate isolation from the impressive developments in descriptive linguistics, which anthropologists either claim as their own or respect with more than a nodding acquaintance.

The interest of psychologists has recently been aroused by provocative analyses, such as those by Benjamin Whorf and Dorothy Lee, which purport to show an intimate relation between the structure of a language and the implicit categories of perception and thought shared among its speakers. Psychologists are properly wary of the conclusions of these studies, based as they are on internal analysis of language structure. The linguist's analysis needs to be matched against direct and adequate data on perceptual and cognitive processes, collected by methods as independent of the structure of the particular language as possible. Psychologists are peculiarly qualified to work on this end of the problem, but to do so they need much more sophistication in linguistics than they are likely to have at present. This so-called "Weltanschauung" problem is one of many on which the psychologist and linguist could profitably collaborate. Happily there are indications that promising lines of collaboration are developing.[23]

Emphasis on More Differentiated Concepts. While our discussion has centered around the broad concepts of personality and culture, psychologists and anthropologists are less likely to talk at cross purposes and more likely to contribute to a common theoretical framework when both employ more differentiated concepts that are better suited to the analysis of process. The psychologist concerned with the discrepancy between what people say and what they do, or with the contrast between unconscious and symbolically formulated behavior, can relate his problems to the anthropologist's distinctions between ideal vs. real, overt vs. covert, implicit vs. explicit patterns. Differentiated concepts of social structure, particularly those of position and role, have been left for treatment in the chapter on "Sociology and Psychology." But these concepts are rooted as deeply in anthropological as in sociological writings, and as Newcomb[24] and Parsons[25] have argued most cogently, their significance for social

[23] Cf. John B. Carroll et al., *Report and Recommendations of the Interdisciplinary Summer Seminar in Psychology and Linguistics,* Ithaca, N.Y.: Cornell University, 1951. (mimeo.) See also H. Hoijer's chapter, "The Relation of Language to Culture" in *Anthropology Today,* A. L. Kroeber, ed., Chicago: University of Chicago Press, 1953.

[24] Theodore M. Newcomb, *Social Psychology,* New York: Dryden, 1950.

[25] Talcott Parsons and Edward A. Shils, eds., *Toward a General Theory of Action,* Cambridge, Mass.: Harvard University Press, 1951.

psychology and personality theory can hardly be overestimated. The related concepts of values and norms, about which much terminological confusion still exists, have as yet been very inadequately assimilated into psychology.

The Emergence of Appropriate Models. If psychologists and anthropologists are not merely to cooperate on problems where their interests overlap but are also to contribute to a coherent and more inclusive body of knowledge about social behavior, they must approach their respective data in terms of mutually compatible conceptual models. Historically, incompatibility—or mutual irrelevance —between psychological and socio-cultural models has been a major obstacle precluding effective interdisciplinary collaboration.

The opposing models most hampering in their consequences were psychological individualism on the one hand and social or cultural realism on the other, the latter expressed in such metaphors as the superorganic or the group mind. Theory in social psychology was long torn between stress on the isolated individual, who somehow had to be related to others to form a society, and on a reified society or group, the basis of which in individual psychological processes then became most mysterious. Since social psychology was committed by its self-imposed charter to concern itself with the bridge between individual and group concepts, the gyrations taken by its developing theory reflect tensions generated by an impossible task. Anthropology was more happily situated, free as it was from commitment to any such bridge-building operation. But similar models offered themselves, and those who adopted the attractive superorganic model of culture as an entity *sui generis* had little occasion to concern themselves with psychological matters.

One of the most hopeful features in the current situation, therefore, is the emergence from diverse quarters of incipient consensus on a model that rejects the sterile dichotomy of isolated individual vs. disembodied group. The model is by no means new; Cooley and G. H. Mead had the essential insights early in the century. But the tradition springing from Mead and Cooley tended to remain the private property of sociologists, in whose hands it retained a speculative flavor. What is new today is that the insights are being rediscovered on all sides, and—more important—they are being employed, refined, and tested as never before in empirical and experimental research.

This new-old model has many variants in the hands of different theorists. But its minimal features can be quickly delineated. It takes its start not from individuals or socio-cultural entities but from the interactions of persons; hence it is sometimes called "interactionism." In any functioning social group, according to this view, the isolated individual is a misleading artifact. Persons achieve communication and avoid randomness in their relationships because each already embodies much of the socio-cultural system in microcosm. That is, the symbol-systems, beliefs, and expectations of one another shared by members of a social group, the modified or emergent motives, aspirations, and standards of evaluation learned in group experience, so transform man as a biological entity that he exists always in implicit relation to others. The notion of men as social atoms somehow to be brought into significant relation to one another is simply a myth, except when applied to collections of infants where, indeed, it is not inaccurate.

With the demise of the isolated individual, the case for social realism—the "group mind"—loses its pertinence. More, to be sure, than the sum of its *individual* members, a group is comprised of interacting persons in their relationships without the necessity for any *tertium quid.*

So much for the synchronic aspect of the model—its pattern for describing social behavior at a given point of time. Diachronically, the model suggests that in the process of social interaction, the actor learns through time to become the sort of person who is capable of the orderly social relationships in which we later encounter him. Within the framework of the model, therefore, the conditions and consequences of the socialization process in its detail become a central problem for research. Through socialization, culture, which at the outset of the life-cycle is exterior to the person and constraining upon him as Durkheim would have it, becomes internalized—inextricably incorporated in his very make-up.

The implications of this model have as yet to be worked out satisfactorily in psychology, particularly in relation to the theory of motivation, where an unquestionable biological bias has resisted conceptual incorporation in the interactionist framework. Developments in this sphere should be particularly important for their bearing on anthropological problems. They must come, however, from theoretically oriented research, not from armchair speculation and model-building.

If these indeed bear some resemblance to the lines along which a comprehensive theoretical framework for the sciences of social man is to be erected—to hazard a most speculative guess—the need will persist for specialized inquiry from perspectives centered not too differently from the psychology and anthropology of today. Process and patterning at both the personal and cultural levels will continue to require special study, anthropology offering to psychology a detailed account of the social context of behavior, and psychology reciprocating with a corresponding account of the interpersonal processes underlying culture. If the comparative method ceases to be the monopoly of cultural anthropologists and modern society the privileged territory of sociologists, the boundaries between these two disciplines should become progressively indistinct. Workers within all three disciplines can perhaps best contribute to the emergence of a comprehensive structure by giving explicit attention to problems of articulation in framing their working concepts and hypotheses.

CHAPTER 4

Psychology and Sociology

TALCOTT PARSONS

THIS PAPER WILL BE concerned with the relations between psychology and sociology as theoretical disciplines, but from a very specific point of view which should be made clear and explicit at the outset. It is written by a sociologist in an attempt not only or even mainly to answer the question of what have been the contributions of psychology to sociology, but rather to attempt to state clearly a framework in which the question of the future fruitfulness of the relations between the two disciplines from the sociological point of view can be worked out. The central question then, is what are the conditions of an optimum "fit" between two theoretical schemes which can make the one as fruitful as possible for the other. The sociological perspective in which these questions are discussed will inevitably entail some criticism of past and partly of present trends of psychology, but rather less of sociology. If it were written by a psychologist about sociology the reverse would be expected. Hence the reader should keep clearly in mind that the purpose of the paper is not a general evaluation of psychological theory, but rather an evaluation of different trends for this specific purpose. How important this function of psychology relative to others may be is a question which cannot be dealt with here.

I

When one speaks of "psychology" and "sociology" there is a certain abstraction involved. Both are rapidly developing disciplines with diverse trends of thought to be found within them. No one writer can speak for his whole profession. But the "personal" element might enter in different ways of which I would like to distinguish two. One might, in such a paper, attempt a critical discussion of the major trends of sociological theorizing going on, and then attempt to relate the problem of the place of psychology to each. On the other hand, one might take one position which, whatever the question of its typicality, is at least by contrast with psychology, clearly sociological, and discuss the whole problem from that vantage point. Because of considerations of space, and because of my greater familiarity with the problems of the particular type of sociological theory with which I have personally been working, I have chosen the latter course in this paper.[1] The reader should hence be aware that a sociologist who thinks in different terms might well see some of the problems of his relation to psychology somewhat differently. The title of the paper is thus elliptical; the full form would be something like "Some problems of the relation of psychology and sociology from the point of view of *one kind* of sociological theory."

However its boundaries may be defined in other respects, sociology is clearly concerned with the observation and analysis of human social behavior, that is, the interaction of pluralities of human beings, the forms their relationships take, and a variety of the conditions and determinants of these forms and of changes in them. The psychologist is traditionally concerned with the behavior of "the individual" though a very large part of the behavior of individuals occurs in relationships with other individuals. Sometimes there is of course even more overlap as when "social psychologists" concern themselves with the behavior of crowds, the formation of public opinion, and the like. Here quite clearly a distinction, if it can be made, must be stated, not in terms of the different concrete phenomena studied, but of a basic abstraction from, or mode of analysis of, data concerning these phenomena.[2]

From the present point of view the focus of sociological theory is held to be on certain aspects of the structure of and processes in

---

[1] Cf. Talcott Parsons and Edward A. Shils, eds., *Toward a General Theory of Action,* Cambridge: Harvard University Press, 1951; Parsons, *The Social System,* Glencoe: Free Press, 1951; and Parsons, Bales, and Shils, *Working Papers in the Theory of Action,* Free Press, 1953.

[2] Thus to say in the concrete sense that public opinion was a subject-matter for psychology but not for sociology would be essentially to say that sociology ought not to study social interaction which in turn would be tantamount to saying it ought not to exist as a distinct discipline.

social systems. A social system in turn I define as the system constituted by the interaction of a plurality of human beings, directly or
indirectly, with each other. Psychology, on the other hand, I hold to
be concerned first with certain elementary processes of behavior, like
learning and cognition, which, however much they may be concretely
involved in social interaction, can be isolated from its processes for
special study, and secondly with the organization of the components
of behavior to constitute the personality of the individual as a system, the system of behavior of a single specific living organism.[3]

This way of defining the relations of the two theoretical disciplines has certain implications which should be made explicit. Their
common reference is behavior.[4] But it is behavior studied and analyzed in terms of a common frame of reference which some of us
have called that of "action." It focuses attention upon and categorizes
the behavior of the organism, and not its internal structure and processes. Behavior or action in this sense is a mode of relation between
an "actor," i.e., an organism or a socially organized collectivity, and
a situation, which may be conceived as a system of objects, of which
the most important are "social objects," i.e., other actors. The action
frame of reference thus leads very directly to the conception of
social interaction. It is the relations between the organization of the
components of action-interaction around the individual organism as
actor on the one hand, the system constituted by the interaction of
a plurality of actors on the other, which is the center of the problems
of the present paper. The fundamental postulate from which the
analysis proceeds is that these are independent and not mutually
"reducible" system-references. Put a little differently, the common
sense of the psychologist tends to hold that, if action is accepted as a
frame of reference at all, it concerns the action of individuals (organisms); interaction is then a resultant which should be accounted
for by extrapolation of our knowledge of the action of individuals.
The common sense of some sociologists, on the other hand, tends to

---

[3] This definition is formulated with reference to the problem of locating the
theoretical center of gravity of psychology within the family of sciences of action.
In no way does it pretend to describe the total field of interest of members of
the psychological profession. In particular it does not locate "social psychology."
The latter I conceive as best defined as an "interstitial" discipline *between* psychology and sociology, thus analogous to biochemistry as between chemistry and
physiology. Cf. my *Social System*, Ch. XII, for a fuller discussion of these problems.

[4] By far the most important case for us is human behavior but neither need
be confined to the human case.

suggest that interaction as such constitutes a system which is over and above and may even have priority over the action of individuals. The contention of the present discussion is that both are right in that both constitute authentically important and independent systems, but neither has priority over the other, *neither provides the premises from which the major characteristics of the other or of action in general can be derived.* Or, it may be put, each provides some of the premises of a general theory of action.

Part of the historic difficulty of reconciling these two points of view has rested on the tendency for both sides of the controversy to set the individual over against society and to identify the concept society with that of social system. This can be very misleading indeed in that it obscures the fact that *any* process of interaction of human beings may constitute a social system. A committee, a work group, or even a family clearly do not constitute, in the usual sense, societies. But equally clearly they are, for the purposes of sociological theory, social systems. A society, then is not only "a" social system, which of course it is, but is also a very complex network of interlocking and interdependent subsystems, each of which is equally authentically a social system. This is the perspective in which I wish to treat the problem of the relations between personality and social system.[5]

One implication of this perspective emerges immediately. If the problem is that of individual as over against society, it is very easy to think that the "unit" of the society is the individual. If, however, one is concerned with the subsystem, with what is sometimes treated as "the group," then the concrete total individual cannot be the unit, because of the simple fact of multiple participations or memberships. It is the role or status-role of an individual which becomes the unit of group, i.e., of social system structure. Simple and obvious as this consideration seems to be, taking systematic account of it alters the traditional perspective on the personality-social system problem quite fundamentally. To draw the most important inferences from this starting point will be one of the central tasks of this paper.

One further aspect of the general action frame of reference needs, however, to be briefly discussed before proceeding farther. Action, it has been stated above, is a mode of relation between a living

[5] The psychological analogue of social system therefore should be "motivational system" or some such concept rather than "personality" which is analogous to "society."

organism and a set of objects in its environment or situation. From this, we may say that, within the frame of reference, it follows that the primary significance of objects in action lies in what comes to be their meaning to an actor. There are many shadings and aspects of meaning in this connection. But primary concern here is with the symbolic levels of meaning. This may be taken to imply that meanings are not "particularized" to the utmost, but come to be organized in systems. Then a given specific object in the situation of action is significant, i.e., "has meaning," because of the way in which it fits into the organized patterning of the meaning-system, rather than simply because of its immediate and isolated impact. This is what we are saying when we refer to its meaning as "symbolic." Further, by virtue of such relations, objects can be connected with each other in complexes of meaning, so that one object in a complex can come to "stand for" others or for the complex as a whole, that is, to symbolize other objects.

The patterned and reciprocal organization of the meanings of objects is what is distinctive about the structure of systems of action; it is this by virtue of which "orientation" to objects becomes determinately stabilized. This is what we mean by saying that action is "culturally" organized, that in a personality considered as a system there is an *internalized* culture, while in a social system the counterpart of internalization in the personality is *institutionalization*. Culture is thus in certain respects analytically independent of its "embodiment" in systems of action, first in that it may be abstracted analytically from actual behavior and treated as a complex of pattern-systems, and secondly in that it can be transmitted from one system of action to another, between personalities by learning, between social systems by diffusion. It is therefore necessary to add the cultural aspect or "dimension" to those of social system and personality in order to complete the frame of reference for analysis of interactive behavior in action terms.

On the basis of these assumptions it is now possible to say something about the nature of the articulation between personalities as systems and social systems, which can serve as a guide to analysis of the theoretical relationships of the two disciplines of psychological and sociological theory. The two types of system are here conceived, not only as interdependent, but as *interpenetrating* in a specific sense. Every social system, i.e., system of interaction of a plurality of individuals, involves a sector of the behavior and thus of the personality of each of the component actors. For purposes of conceptualizing the social system, this is conceived as a role which, within the range of situations defined by his membership in the group or interactive system over a sufficiently long period of time, is from one point of view a series of expected or patterned behaviors, not of just one type, but a pattern of types varying according to the development of the inter-active situation. These will include phases in which the individual is not actively participating in the activities of this particular group, as when, at home and away from his job, a man does not interact with his colleagues. Nevertheless, his membership in the work group has not ceased to be important to or part of his personality. This we would call the phase of "latency" of his job role.

This participation is organized and structured, it is not random activity. As part of the personality system it has to be motivated in the sense that the pattern of activity is both regularized or stabilized so that it is not interfered with by other elements and it must be re-sponsive to the interactive situation as it develops, especially that is to the acts of other members of the interactive system. Ego's "per-formances" then are interdependent with alter's "sanctions" and this interdependence is what we mean by the process of the interactive system.

At the same time, each of the other members of the interactive system or group is an object to ego, as indeed he, in this role (as in others), is to himself. Each has qualities of which status in the group is one of the most important aspects. In this aspect, each object in the group has meaning to ego, it is a symbol or a complex of symbols. The mutuality or complementarity of orientations then means that the interactive system as a system must, as a condition of any rela-tively stable state, have a determinate patterning of complementary meanings of objects and orientations. This relatively stable pattern-ing of meanings is what we mean by the "common culture" of the interactive system.

The necessity and importance of a common culture for an inter-active system does not imply that it is "static," that "nothing hap-pens" or that change of state is precluded. It means only that the particularities of each act and each changing situation are not ulti-mately determinant of process but rather that the process is *or-ganized* relative to these particularities and that, within the action frame of reference, the relevant meaning of the concept organization involves the patterning of symbol-meaning relationships. At the same time the interactive system as a system cannot be solely determined by these meaning-patterns, since it is subject to adaptive and integrative exigencies, i.e., to conditions of the nature of the situations and of the actor-units of which it is composed. As a resultant then of its cultural patternings and of the situational and integrative exigencies of the system and, finally, of the motivational forces involved, the interaction system at any given time has a determinate structure. It has parts—the role-units—which stand in relatively determinate relations to each other as objects of orientation, as performing entities, and as sources of sanctions.

Now it has been noted above that social interaction system and personality system interpenetrate. On the more microscopic level where the relevant units are roles of individual actors rather than of collectivities, the role-unit of the interaction system is in fact a sector of the personality as a system. Because of this interpenetration of the two systems, their interdependence must have certain special features, must, that is, be subject to certain constraints. Because of the fact that as different systems they are subject to different sets of adaptive and integrative exigencies, we may say that the focus of these constraints rests on the presence of the common culture. The patterns of symbol-meaning, that is to say, which are constitutive of the structure of an interaction system, must in a stable state also be constitutive of the personality systems which interpenetrate with it. The common culture must extend *into* the personalities constituent of the interaction system, not merely come "up to their boundaries." This is the meaning of Durkheim's aphorism "society exists only in the minds of individuals."

The nature of the independence of the personality system relative to social systems may now be more clearly seen. For each individual the living organism is unique and individual in two respects. In the first place it is the source of the motivational energy of his action which as such cannot be shared with any other. Then secondly his body as an object is a unique facility and set of reward-objects. It has qualities and performance-capacities with reference to which he holds a natural monopoly. In part these features of his body serve to categorize him with others, as by sex or age or intelligence, but also very much to differentiate him from others. In this connection it should not be forgotten that the physical location of a person's body imposes very specific conditions on his action; e.g., he can, if resident in Boston, attend a conference in New York only if he is physically transported from one location to the other. In both these fundamental respects each personality is unique, i.e., as a system independent of any other, because each organism is a distinct, boundary-maintaining system.

But there is still a third fundamental source of the independence of the personality as a system. This derives simply from the fact of role-participations in social systems. In any given system of social interaction in the nature of the case no two participants can be in exactly the same role, precisely because such systems are differentiated systems. This means that categorization by one member of himself as an object in relation to other objects must be differentiated from that of the other participants. Their relations to each other can be identical only in the limiting case of a perfectly symmetrical system. A second aspect of participation is an inference from the fact that a society is a complex web of subsystems of social interraction, namely, that a given individual in some sense participates in a unique combination of such subsystems. Thus while in our society both husband and wife participate in the family of procreation, though in differentiated roles, the wife does not ordinarily participate, except in a peripheral role, in the occupational interaction system of the husband. Conversely, each of the husband's male occupational associates participates in a different family system. The structure of such role-participations will vary from one society to another, but the basic fact of differential participation is a fundamental of social structure with profound implications for the theory of personality. Finally, the above two sources of differentiation of personalities, relative to participation in social systems, are compounded by a third, namely the fact that there is differentiation, social-participation-wise, by life-history. Some patterns of succession by stages of the life cycle are highly standardized. But others allow for much variation so that the cumulative results of previous role-participations serve to differentiate individuals rather than to assimilate them to standard types.

II

The above considerations should be sufficient to counteract the common misunderstanding that emphasis on sociological factors is a threat to the analytical independence of the personality level of analysis, or to the ideas of either the "uniqueness" or the "autonomy" of the individual personality. On the contrary it can be argued with considerable force that such considerations help a great deal to give a firm foundation to our intuitively perceived insights and to psychologically established knowledge in this field.

This is true and extremely important. But it is secondary to the basic problem of this paper which is, from the sociological point of view, to state some of the most essential requirements of a psychological theory which is maximally useful to the sociologist. We may now, on the background of the general considerations which have been reviewed, attempt to make this problem somewhat more explicit and specific by sketching two important problem-areas in which the two disciplines meet, where, that is to say, important sociological problems are involved, but considerable use must also be made of psychology.

The first of these which I would like to consider is the problem of "social control" in those cases where the "vicious circles" of motivation to deviant behavior have become firmly established using the case of illness in relation to therapy for illustration. In so far as states of illness can be considered as motivated, either etiologically or with reference to resistance to recuperative processes, or both, and the motivation is unconscious in the sense that getting or staying sick cannot simply be treated as malingering, we may speak of it as deviant behavior since in a broad way, to be in good health, physical and/or mental, so far as it is possible, is expected in terms of the values of our society. The treatment of illness as deviant behavior has proved illuminating partly because it has been possible to establish a broad continuity between it and other types of deviance, such as delinquency.

Leaving aside the problems of the manner of genesis of motivated illness in the processes of social interaction and in the personality of the individual, we may raise a few of the problems posed by the process of therapy. Whether the difficulty be "psychosomatic" or "mental" there seems to be very widespread agreement among the relevant professional groups that it is usually very difficult if not impossible for the individual to "cure himself," and also that interaction in his ordinary social relationships, in family and job, will usually not automatically "cure" him. For the sociologist this is to say that effective therapy is, among other things, dependent on the establishment of a special type of social relationship. From "being sick" the individual comes to be a "patient"; he comes to play a particular type of role in a particular type of system of social interaction.

From the sociological point of view, the "sick role," the role of therapist—in the most important type of case a physician—and the role of patient of a therapist, are all integral parts of the structure of the social system, and each of these roles and the interaction between them have distinctive sociological characteristics. Thus illness as a social role, as distinct from merely a "condition," involves first an exemption from normal social obligations, and second a claim to be "helped" in "getting well," both of which have to be socially legitimized. Not just everyone who says he is unable to work because he "feels bad," for instance, is treated as sick. Secondly this legitimation is partial and conditional. By being categorized in the society as sickness, the state is evaluated as in itself undesirable, and an obligation is imposed to try to get well as expeditiously as possible. Seeking therapeutic help is then virtually an obligation, and once a therapeutic relationship has been established, the patient assumes the obligation to cooperate to the best of his ability with his therapist.

Matters are similar on the other side. The role of therapist is partly defined by specific technical competence, validated by formal training. But this fits in a sociologically distinctive framework, in our society typically in that of the "professional" role, which imposes a pattern of behavior on the therapist which cannot be simply deduced from the content of his technical competence; for example, his obligation to "help" his patient and be concerned for his welfare as distinguished from the typical business or commercial pattern of regarding him as fair game for financial exploitation.[6]

Another set of considerations about the therapeutic role has recently emerged as highly important. If illness be conceived as a form of deviant behavior, then the therapist stands in an interesting "interstitial" position. In his permissiveness and supportiveness, and in the confidential intimacy of his relation to the patient, he partially and conditionally participates in the world of the sick. He "understands" how his patient feels, and to a point legitimizes the latter's orientation; he permits its expression without punishment and does not withdraw support because of the deviant character of the sick person's ideas and wishes. But this is only a conditional acceptance.

[6] The relevant features of the role of illness and of the therapist have been described and analyzed elsewhere. Cf. Parsons, *The Social System*, Ch. VII and X.

The therapist also, and progressively more so as therapy proceeds, represents the society of the healthy. Particularly in analysis of transference reactions he shows his patient how deviant his reactions are, and on occasion rewards him by approval for his insight into his own motivations. The deviance is accepted only provisionally and in order to help the patient overcome it. The latter would, it seems, not be possible if the therapist participated only in the world of the sick, if, that is, he were not himself defined as well, and did not accept the normative standards of the healthy world, not only for himself, but eventually for his patient.

These basic features of the social roles of the sick and of the therapist, and of the nature of their interaction are, as I said, parts of the social system. They cannot be explained in psychological terms in the sense which would involve ignoring the fact that they are institutionalized as parts of the social system except as the "fact" to be explained. Furthermore they are clearly of first rate empirical importance in explaining in part how and why psychotherapy works as well as it does in modern Western society.

But the sociologist who lets it go at that, who simply by-passes the problems of the mechanisms of personality process by which, under the conditions of institutionalized (not in the hospital sense necessarily) therapy of a person socially categorized as "sick," a certain order of effects come about, is leaving a vital aspect of the empirical problem areas hanging in the air. With empirical validation from statistical sources he may, with due qualification for imperfections of available evidence, be able to establish that and even how well it works, but at best in a very limited sense how it works. To get farther he must have knowledge of the operation of the personality as a system, of why the various relevant features of the therapeutic interaction system are important as conditions of operation of certain psychological mechanisms, including what the psychological effects would be if the conditions were altered in specific ways.

But for this to work out most advantageously from the psychosociological point of view, it is necessary that treatment of psychological mechanisms in the personality system should, so far as possible, be articulated with the categories of the social role-system in which the interaction takes place. Thus in the current case the problem for the personality of the patient is socially posed in terms of the categories "sickness" and "health." The psychological tendency has been to take these for granted as obvious to common sense. But this is a dangerous assumption. Above all, illness in the aspects important to the present argument is not simply a disturbance in the organism which biological criteria suffice to specify and interpret, but is a form of behavior relevant to the system of social interaction. Above all, it must be understood in terms of the common values of a system of social interaction as a central point of reference. This common value system is in turn conceived to be internalized in the personality; one can only speak of sickness in this sense if the deviance is "unconscious"—otherwise it would be malingering—and hence involves personality conflicts organized about the conformity-deviance axis.

That this basic point of reference has in fact entered into personality psychology is above all evident in the emergence of the concept of the superego in Freud's work, and its connection with guilt and shame as motivational categories of personality psychology. It is from starting points such as these that the articulation of personality psychology and sociological theory must work. The conception of either "mental" or "psychosomatic" illness as a "condition" independent of social interaction does not do justice either to the significance of the state of the personality as such or to the problem of analyzing the processes by which effective therapy may be possible. At the same time, taking account of the social system aspects, far from eliminating the relevance of psychological analysis, highlights its importance by giving it a particularly sharp focus.

A second example concerns certain aspects of the problems of the socialization of the child with which I personally have been working.[7] The problem is that of how to predict from certain features of the role structure of the family in which a child—in this case a boy—is socialized, what level and what qualitative type of adult occupational role he is likely to attain. The ground for attributing substantial significance to aspects of personality internalized in the family has been cleared by showing that, when broad family socioeconomic status and ability (measured by I.Q.) have been taken into account there is a substantial residual variance and, secondly, by showing that the differences between those boys, for instance, of relatively high ability but low family status who do and do not go to college, cannot adequately be accounted for simply on grounds of access or lack of it to economic resources. The problem, then, is by

[7] In connection with a study of social mobility in collaboration with Samuel A. Stouffer and Florence R. Kluckhohn.

analysis of socialization as a process, to predict from (or conversely explain in terms of) the structure of his particular family as a social system, the outcome of a selective process in the allocation of social roles, by showing what are the kinds of personality which, on the one hand are produced in the socialization process, on the other are "predisposed" to assume certain types of occupational roles.

Consideration of the problem has made clear that one cannot get very far by what might be called a simple "correlational" approach to it. One might take available "census-type" classifications of occupational roles, and current psychological categorizations of personality types, such as introverted vs. extroverted, impunitive, intropunitive, extrapunitive, etc. and attempt to correlate them. On the other hand one might take sociologically common sense classifications of family type, with reference to number of children, birth order, father- or mother-dominated, "authoritarian" vs. "democratic," etc. Though some of these are undoubtedly significant, particularly birth order, it seems this as such is not a really fundamental attack on the problem.

The procedure being followed is different. First we have attempted to work out a classification of occupational roles as a reference system in terms which go substantially beyond those yet current in sociology. Above all an effort has been made to derive this as far as possible from considerations of general theory so as to maximize the chances of its connecting up with other analyses. This has necessitated extensive analysis, particularly into the structure of social stratification.[8]

The next task has been to work out an analysis of family structures which could treat variations in terms directly comparable with those used to delineate the range of variability of occupational roles. Both these tasks, it will be noted, are essentially tasks of the sociological level of analysis of the structure of social systems. Basic orientation on these levels is, we believe, essential even to statement of the problems of central concern on the psychological levels, if that is to say, psychological theory is to be useful in solving the original prob-

[8] Two very preliminary statements of results of this work have been published, first a brief discussion of the occupational reference system as Ch. V, sec. viii of Parsons, Bales, and Shils, *Working Papers in the Theory of Action*, and second the paper "A Revised Analytical Approach to the Theory of Social Stratification" in Bendix and Lipset, eds., *Reader in Social Stratification*, Glencoe: The Free Press, 1953. Empirical research involving these categories is not yet ready for publication.

lems of why some boys elect one type of occupational role (and qualify for it) while others go another way.

In the course of this work, however, analysis of the structure of the family has led almost immediately into analysis of the process of socialization. The key to our approach here has been the recognition, first worked out in detail in the occupational field, that a social system is a complex network of subsystems interlocking with each other. This proved to be true even of so "simple" a unit as the American urban family. In particular it has been important to treat the mother-child relationship (in the preoedipal stage) as a distinct subsystem. The child, then, may be conceived as coming to be progressively integrated in a series of successively wider and more complex systems of social interaction, first the mother-child system, secondly the family as a system, and third the community outside the family—which of course also needs in turn to be further broken down.

This conception of socialization as involving integration in a series of systems of social interaction, the structure of each of which has been analyzed in technical sociological terms, has made it possible to treat the process of internalization of cultural patterns in an orderly, and we think, psychologically acceptable fashion, and in a way which analyzes the motivational systems and subsystems of personality in terms directly cognate with the analysis of role-expectations in the relevant social systems, including the future occupational system.

With respect to the structure of the interaction systems, to the nature of the learning process and, finally, to the structure of personality, certain relatively clear patterns have already emerged. Let us review each of these in turn.

First, with respect to the systems of social interaction in which the child becomes involved, we can think of a process of successive differentiation. The simplest state sociologically is that often called by psychoanalysts and other child psychologists "oral dependency," in which the child is, in Freud's original sense "identified" with the mother, in that he has not yet differentiated his own "ego" from the "mother-child identity." A second stage of differentiation is reached when the child develops an "autonomous" need to love as well as be loved by the mother, and with this develops a conception of self as "loving and being loved" and not merely as dependent. From the sociological point of view this is a more complex system, though it still involves only two "natural persons," mother and child.

The conjugal or nuclear family as a system represents a further stage of structural differentiation. The number of roles in it is naturally a function of the number of persons. But the structural ordering of the roles, on a broad basis, must be dominated by two major axes of differentiation, which in common-sense sociological terms are generation—parents vs. children—and sex. On this basis there are four and only four generically basic types of role in the nuclear family, parent of each sex and child of each sex. The child by ascriptive inevitability has, on this basis, the "choice" of only two roles since only over a long period of time can he hope to become a parent or a member of the older and hence superordinate generation.

We therefore conceive of sex-categorization as the decisive aspect of emerging from relatively exclusive relationship to the mother into full membership in the family. Then the father becomes significant as differentiated from the mother and the sibling of opposite sex becomes significant as differentiated from ego.

The conjugal family in this sense is, we have reason to believe, a prototype of a social system in that it incorporates what we believe to be the two most elementary axes of differentiation of social structure, namely, "power" or "prestige"—whichever way one wishes to look at it—and primacy of "instrumental" function for the system and of "expressive" function. The child who has come to be a full-fledged member of a nuclear family may thus be regarded as oriented to some of the fundamental facts of life. He must recognize that there are these basic types of role and that, at any given time, he must locate himself in the system composed of them in the sense that he must recognize his own role as part of the system and hence as differentiated from but also related to those of the other members.

But the series cannot stop at the boundaries of the conjugal family. This in turn is a subsystem of the wider community structure in which the family is embedded. I shall not take space to carry the analysis farther here, but only say that in addition to, not in place of, his (changing) role in the family a boy acquires roles in school and peer group, so that in "latency" and adolescence he has three main interdependent foci of his life (in our society of course). Only after that does he marry and by establishing a new conjugal family greatly attenuate his ties to the old. In terms of social structure the systems in which he participates are, at each stage, as a total set, progressively more complex.

This is the sociological setting in which the psychological problems concerning the processes of personality development and the nature of the resulting structures of personality have to be raised. But even here our treatment has been guided by sociological considerations. Recent work has strongly suggested, if not established, a relationship between the succession of phases found in the processes of interaction in small groups, and the order of the therapeutic process and hence more generally of operation of mechanisms of social control. This is that the order of phases of the therapeutic process is the reverse of that found in ordinary "task-oriented" groups.[9] Then it appeared that the Freudian paradigm of the main stages of the psycho-sexual development of the child could be fitted into the same reverse order of phase succession.

To spell out the above suggestion briefly, let us look at the four stages of development and then the four stages of therapy to show how, in a sense, they do reverse the order of normal task-oriented group process. In development, the four stages we recognize are these: (1) maternal care is focused on the gratification of organic needs; the mother is permissive with respect to all organic tensions; (2) there is the establishment of an active love-attachment between mother and child; (3) there is a stage when disciplines become more prominent—the child is denied reciprocity for both dependency and aggressive impulses, and his positive achievements are selectively rewarded; (4) finally the child achieves emancipation from primary integration in his family of orientation, becoming primarily sensitive to rewards and punishments outside leading up to the establishment of his own family of procreation, and, if male, of his occupational role. In therapy we have roughly the same sequence of stages: (1) there is a stage of permissiveness to relieve the patient's fear; (2) there is a stage of transference during which in response to the therapist's supportive attitudes an attachment of patient to therapist is built up; (3) the therapist, by "interpretations" begins to manipulate rewards making them conditional upon the patient's adaptive responses to therapy; (4) the patient is weaned away from therapy and sent back to face the adaptive exigencies of the world.

In normal task-oriented groups, on the other hand, an almost exact reversal of this order has been observed by Bales: (1) at the outset the group is concerned with the adaptive exigencies surrounding its problem, the search is for information as to the nature of the problem and the nature of the environment; (2) next there is a stage

---

[9] These considerations are discussed in *Working Papers*, Ch. V, sec. vii.

when instrumental action is undertaken conditioned on the best guess as to what overt actions will lead to reward; (3) after the group's concrete problem has been solved, there is a stage when the solidarities disrupted by concrete action are resolidified; (4) after disintegrative tendencies have been relieved in such a fashion, the group structure becomes latent and there is an interval of permissiveness to relieve other motives that have been held in check.

It is, of course, a commonplace that socialization is a complex of processes of learning. We are applying this knowledge with the addition of two very important considerations which serve to organize and codify, for our purposes, the immense body of available knowledge of learning processes. We treat learning as a response to changing features of the situation in which the child is placed, most strictly, changes in the child-social situation system as a system, which include biological maturation but also other factors, above all the responses of other people to that maturation. Within this framework we think of that process as involving shifting balances between a plurality of factors, and secondly as involving certain elements of discontinuity, so that it is not a linear process, but one which goes through a kind of a spiral of phase-cycles.

On the level closest to the kind of social interaction involved in therapy, we think of the four aspects, previously analyzed, of permissiveness, support, denial of reciprocity, and manipulation of rewards, as all involved at any time, but as having relative predominance in that order for each main phase cycle. This may be illustrated by one main cycle, that involving the surmounting of the oedipal crisis. The relatively stable starting point, we may say, is the child's love-attachment to the mother. In the security of this attachment he is, even after the "pressure" begins, given a certain permissiveness, this time for expression of his dependency needs in this relation. He is further supported by the steadiness and relative unconditionality of the mother's love for him. However, in gratifying his needs for dependency and security he is emboldened to attempt unacceptable overtures to mother and father, many of which, above all those involving aggression and various aspects of dependency, are rebuffed. Certain performances, however, which are judged to be in accord with the requisite levels of maturity, are rewarded, above all with attitudes of approval, and tend to be learned while the unacceptable impulses tend either to be extinguished or to be repressed.

We cannot take space to spell out these processes in full, even so far as this now seems possible. Suffice it to say that knowledge of classical and instrumental conditioning processes, and conditions of cognitive learning such as discrimination and generalization can be effectively built into such an analysis and help to build the bridges between the various stages. In other words, a great deal of the learning theory which has developed in connection with experimental animal psychology has proved usable in this connection.

Secondly a great deal of psychoanalytic theory has also been found to fit in. The stages of psychosexual development have already been mentioned. But equally gratifying is the use which can be made of the psychoanalytic mechanisms, including those of learning and of defense and adjustment. The most important to mention here is that of identification. Freud's original usage has already been mentioned as associated with the "mother-child identity." But further it has been used in relation to what we call the process of internalization of social objects, by which the superego, for instance, is formed.

This brings us to the third set of considerations mentioned above, those concerned with the structure of personality. The most important thing here is a fundamental theoretical guiding idea or insight—rather than a specific hypothesis—which can be regarded as an inference from the view of the interpenetration of personalities and social systems as systems which was stated above. This is that the patterns of organization of personality as a system, as distinguished from the more elementary components of process, motivation, cognition, etc., which go into the organization, are derived from the structure of the successive systems of social interaction into which the individual in question has become integrated—more or less completely, of course. This means that the structure of the personality is a kind of "mirror-image" of the structure of the social object-system. It consists, we may say, of internalized social objects, in the sense that it is a patterned system of the meaning of these objects to the actor.

We should be very careful in the interpretation of these statements. They clearly do not mean that a personality as a system is simply a reflection of the social situation at the time. This would be a negation of the postulate of the independence of the personality system. Or, put a little differently, it would negate the cultural character of the organization of systems of action in the sense in which this involves the transcending of the specific particularity of current situations. The original specific social objects are the "points of reference" or "foci of crystallization" of personality structure. They are generalized from, and past objects are superseded by later ones, e.g., by "substitution." There is hence both generalization involving in a certain sense abstraction and a temporal depth aspect of the structure of the personality. But these considerations do not negate the original propositions, but rather reinforce them, since they are in accord with equally fundamental elements of the general theory of action.

It is this set of propositions which enables us to link the structure of the personality specifically and technically with that of social systems. Put in psychologically relevant terms the main point is that the basis of the link is that the "stimulus-conditions," which have been strategically crucial to the development of personality as a system, are organized as a structure of socially interactive relationships, varying over time, so that learning processes are shaped to conformity with the conditions imposed by the total set of interactive participations of the individual. His lack of fit in one may be balanced by better fit in another of the subsystems. But the basic fact is the dependence of human personality on the experience of social interaction.

There is one further aspect of the patterning of the socialization process which has become progressively clarified in the course of this work. The development of personality should, we feel, in its major structural outline, be treated as a process of the differentiation of an initially simple system into a more complex one. We have been able to advance the idea that the principle by which this differentiation takes place is that of "binary fission" of objects and hence motives in a series of 1–2–4–8–16 etc. What has been called the "mother-child identity" is the starting point of the series, which progresses through the "love-dependency," the "full-family" stage, etc.

To render this more concrete (while not being accurate in all details), we may say that the first world of the child which has no sharply defined regions (the mother-child identity) gradually separates into two regions, one where the parent is present and the other where the parent is absent. The former becomes gradually focused as the child's concept or idea of the parent, the latter becomes the child's idea of the self. Then, cross-cutting this division, a new bifurcation develops between expressive and instrumental regions (of both parent and self). This instrumental-expressive distinction forms the basis, we believe, for later sex categorization: thus the idea of the parent separates into the idea of the mother and the idea of the father; and the idea of the self separates into boy-child and girl-child. Later a further distinction transcending all of these is introduced, dividing each of the four paradigmatic objects (mother, father, brother, sister) again into a particularistic (co-family) and a universalistic (outside-the-family) pair of subtypes. Still further differentiations of course occur but these will suffice for illustration of the principle we have in mind.

This is not the place even to attempt to explain the raison-d'être of such a scheme. Suffice it to say that when the sixteen "cell" or part level is reached, it is possible in terms of structural categories to match the differentiation of "need-dispositions" of the personality with the basic role-structure of a complete society, not merely a specially simple interaction system. This includes the reference system of occupational roles to which attention was given above. We feel, therefore, that the basic theoretical problem of closing the gap, on a sociological level of the analysis of structure, between family as a special subsystem of the society, and occupational system, as another, much more highly differentiated subsystem, has been solved, and that this has been done in such a way as to integrate the sociological analysis very directly with that of the development of personality on a psychological level.

III

In the light of the above discussion I would now like to review very briefly some aspects of four main movements of psychological theory[10] which play an important part in current sociological thought and its background. The four are: the early social psychology associated with the name of McDougall; the "behaviorist" movement, especially in relation to "learning theory"; the Gestalt psychology and the variant of it represented by Kurt Lewin; and finally the psychoanalytic movement. These four trends are not by any means representative of all psychology but they will serve to illustrate the central problems of this analysis of the relations of psychology to sociological theory.

[10] I am grateful to Dr. James Olds for helping me, in discussion, to clarify my understanding of the conceptual structure of several of these psychological theories.

McDougall's instinct theory serves as a convenient starting point because it played a prominent part in the general intellectual situation in the sciences of action in this country a generation ago out of which the problems of this paper have grown. It is perhaps fair to say that it constituted one of the most serious attempts up to that time to formulate a general theory of social behavior.

In the first place it may be said to have been fully aware of the limitations of the rationalistic ideas which had stemmed most directly from the utilitarian tradition and from the dominance in the social sciences of economic theory. McDougall did not of course deny that rational action occurred, but correctly saw that it must be understood in terms of a matrix of non-rational factors. The key concepts of his analysis of the non-rational elements are sentiment and instinct.

McDougall's analysis of the sentiments is the part of his theory which is closest to modern sociological needs. Because of the ways in which components from various instincts are shown to be organized about interests in different objects, this analysis contains many elements of the kind of psychological analysis which is needed now, and has in fact been neglected by other movements in psychology. At the same time, for understandable reasons, McDougall used the concept of instinct as the main organizing focus of his general theory. Essentially, that is to say, he resorted to biological theory for his major frame of reference. In view of the immense prestige of the biological sciences at the time, and of the relatively unsatisfactory state of the sciences of action, this was understandable.

It is also probable that with more recent developments of biology the subject of instinct is likely to have considerable more attention paid to it again than in the recent past. Nevertheless McDougall's emphasis was unfortunate for the general development of psychology since it diverted attention from analysis of the elements of structure in the situation of action by placing the main elements of structure so far as it was relevant to the analysis of social action, in the genetic constitution of the organism. In so doing it ran directly counter to the increasing emphasis on cultural relativity which was beginning to be prominent in anthropology and sociology at just about the time he wrote. This was also very much in line with the failure to see the possibilities of independent analysis of social systems as systems, which was already available in the work of Durkheim and Weber at the time, and also beginning in this country with Mead, Cooley, and a little later, W. I. Thomas. A persistent note in McDougall's writing is reference to psychology as furnishing the foundations of any scientific study of behavior, meaning that a "theory of human nature" underlay any study of how human beings behaved in interaction with each other and development of this was the principal task of psychology.[11]

There is a good deal of discussion of the differences between animal and human behavior, and hence of course concession to the importance of learned elements in human behavior. But the crucial point is not the estimate of this balance, but the source from which the main points of reference for the analysis of the structure of action as system were drawn. In this respect culture and social system quite clearly occupied the second place. The main road to theoretical advance was laid out as that of better knowledge of constitutional factors, and thus the road which has since proved to be the main one of interest to social science was not only not followed up, but clearly given the secondary position. From McDougall's point of view, of course, sociology, where it became relevant at all, would be conceived essentially as social psychology since an autonomous theory of social systems had no place in his scheme.

The behaviorist reaction against instinct theory of this type represented both an advance and a retrogression. It was an advance in the sense that it cleared away positive theoretical views which were in conflict with the needs of integration with sociology and anthropology, and also in that it produced an enormous volume of careful study of elementary processes of behavior. But it was a retrogression in that interest in the problems of the organization of behavior in systems almost dropped out of the picture for a generation. This "elementarism," amounting to the positive dogma that the conditioned reflex or the stimulus-response sequence, was an atom so fundamental that only knowledge of it as such could get us ahead in any field of action, seems to be the focus of emphasis in the move-

[11] McDougall did attempt to face these problems, particularly in *The Group Mind* (1920). Much of this book is surprisingly modern and relevant. It is clear however (1) that he identified "mind" as the subject-matter of psychology with what here we have called "action." (2) Lacking any concept sufficiently close to that of role, he had great difficulties with both the relations of "society" and its subsystems and social systems and personalities. In other words he saw, but did not solve, the crucial problem of the relevance of those aspects of the *structure of the situation* which in social system terms are crucial to the link between the theories of personality and of social system with which this paper is so very largely concerned.

ment. Thorough knowledge of the element was to precede any knowledge of integrated process.

The playing down of the genetically given elements in the structure of behavior, which undoubtedly went too far, had the virtue of focusing attention on the processes of learning, and in proportion to attribution of importance to it, on the possibilities of variability of behavior. This brought psychology distinctly closer to the developing social sciences. Furthermore, as noted above, the elementarism of the new behaviorism happened to coincide with a movement in anthropology, the replacement of the older evolutionary theories (which had affiliations with instinct theory through biology), by a radical "trait atomism" which regarded a culture as a chance collection of randomly assorted traits, the coexistence of which was explicable only in terms of historical accident.

I have emphasized above that the theory of action has proved to be applicable, in both personality and social system aspects, over an indefinite microscopic-macroscopic range. What the behavioristic movement did was to reify a particular sector in this range, that of the S-R unit, and maintain quite illegitimately that only the phenomena as studied on this level had fundamental significance for the theory of action. Logically this is strictly equivalent to saying in mechanics that only processes of terrestrial falling bodies were fundamental, that celestial mechanics was only a "field of application" of the "real" theory of mechanics, which was a theory of the behavior of bodies within a certain range of mass and velocity, over certain specific time spans. This of course is sheer nonsense, the data of personality organization and of social system process are just as authentically fundamental as any other data of human action, and theory relative to them is just as much "basic science" theory as that of elementary animal learning.[12]

This view of the microscopic-macroscopic range of applicability of the theory of action as a conceptual scheme implies that whatever is isolated for study as a system should in principle be regarded as part of—thus a subsystem of—a larger system. This in turn means

---

[12] If anything, psychologists of the elementarist learning theory persuasion have tended to ignore rather than tackle these issues. Though perhaps considered now out of date, the book of F. H. Allport, *Institutional Behavior,* is a good example of the logic of the position which has been implicit throughout. According to Allport, the sociologist who deals with systems of social interaction as independent systems, is guilty of the "institutional fallacy," which only confining theorizing to behavior-element levels can remedy.

that only in a limiting case can the environment or situation of the processes which occur in the isolated system be treated as random relative to the structure and processes of the system of reference. The elementaristic dogma systematically cuts the psychologist off from recognition of this problem. This is, in my opinion, a serious matter for the theory of animal behavior, but on the human cultural level it is no less than fatal. For the organization of behavior comes to be structured, so far as action is cultural and learned, in terms of the meanings of situational objects. But in turn such meanings cannot be simply a random collection of units unrelated to each other; they must be conceived as constituting to some degree a coherent system of meanings. The element of system in meanings comes, however, from the organization of the stimulus situation which, on the socio-cultural levels is a social organization. Only then, by treating the stimulus situation as organized in social system and cultural terms can an adequate theory of the more complex levels of organization of behavior be built up.

The same problem arises on another level, namely, with reference to time. The classical behavioristic experiments produce a well-recognized "drive," e.g., by starving the animal for a considerable period, and then study the behavior under experimentally controlled conditions, leading to the "reduction" of the drive. Even in this case, S-R behaviorism is prone to treat the drive as either another internal stimulus element pushing the animal on to behaviors, or as an activator of responses lowering thresholds generally, but not selecting responses with any present concern for their consequences. Completely ignored by S-R reinforcement psychologists is the function of drives involved in changing the phase or the cathectic significance of internalized objects which are integrated by former learning into some sort of a temporally ordered system (the associated chain).[13] Thus, these psychologists ignore the basic internal organization of the food cycle itself. Even more important from the present point of view the cycle of food-getting activity *in relation* to other motiva-

---

[13] There seems to be an important parallelism between S-R reinforcement theory's neglect of the "up front" motivation, and the neglect of the principle of "inertia" prior to classical mechanics. Both oversights forced theorists to account for activity entirely in terms of the "push" from behind. As soon as we get an internalized object motivating from up front as a part of the personality system itself, we have something of an "inertia," a natural direction in action, and cues only determine the path to be taken, not the general direction of action. On the possibility of using a parallel concept of inertia for the analysis of action cf. Parsons, *Social System*, Ch. IV, and Parsons, Bales, and Shils, *Working Papers*, Chaps. III and V.

tional systems is not studied.[14] A central part of this problem is what is going on in the periods between the reduction of the drive and its reactivation. In the case of genetically given "primary" drives this may be referred to the physiological processes of the organism; but what of the case of "learned" drives, e.g., to complete writing a scientific paper? It is seldom that the task is completed at one sitting. Under what circumstances do other motives intervene, and what is the nature of processes occurring during the "latency" of this particular drive?

In this whole context, it seems to me, another aspect of the structuring of the situation of action is crucial, namely its structuring over time in phases. On the socio-cultural levels this means the relation of any given actor's action to the temporal order of the activities of other actors; it is the *expectation* aspect of social interaction. The problem of the behavior of one individual or one organism can here be treated in one of two ways. One is that he may be conceived as adapting to a situation which is treated as given, but this situation is an organized or structured one in which not only are the objects related to each other but their behavior undergoes more or less orderly processes of change over time, and gearing into this development of the situation over time is an essential aspect of the organization of his behavior. The second way to deal with it is to take two or more individuals in interaction over time. In this latter case one at least of the relevant systems of action becomes a social system. But if we once grant that the situation of behavior is structured over time, and that organisms are mutually sensitive to each other's behavior, then the analysis of behavior cannot dispense with the concept of the social system. In the light of these considerations the effect of elementarism is, by what is really no more than a logical trick, to make it seem plausible that the "fundamentals" of behavior can be worked out with reference to the behavior of the individual organism alone. Then, it is alleged it is possible to introduce the additional conditions given by the possible presence of other organisms, and thereby deduce what will happen in the more complex systems. The fallacy of this procedure should be clear. If what is purported to exist is not simply a theory of artificially isolated S-R sequences, but of the organization of behavior systems and their functioning as systems, then the reasoning is circular. Unless, as in instinct theory with

---

[14] Attempts by reinforcement theorists to cope with the Tolman latent learning problem have been concerned with the interrelations of drives, but in *ad hoo* fashion.

which the behaviorists will have nothing to do, organization is put into the genes, *it must be found in the structure of the situation,* both at a given cross-sectional moment and over time. But this structure is an aspect of the system of interaction in which the individual has participated in the course of his learning experience. It is clearly not legitimate to explain this system in terms of itself. In general we may say that behaviorist social psychology has never transcended this dilemma. On the psychological level, to do so requires fulfillment of two conditions. The first is the development of a coherent theory of the behavior of an organism as a system, not merely of the elementary S-R unit of behavior. The second is the explicit theoretical categorization of the structure of the "stimulus-situation" as independent of the "mechanisms" of the learning process. But on socio-cultural levels, this categorization is an aspect of the theory of the social system. The behaviorist attempt to derive the theory of social systems from the elementary theory of behavior rests on nothing more than concealed circularity of reasoning.

Several developments following the advent of Gestalt psychology point at different fruitful beginnings of study of psychological systems; but there is in each of these developments a tendency to press the emphasis on particular dimensions of systematization while never quite reaching the point of systematic integration of the various different dimensions. The foci of systematization which we find within the work of the Gestalt and related schools are these: the perceptual object, the perceptual field, the memory trace, the sign-gestalt.

The perceptual object is the original system-concept of the Gestalt school; their position in a nutshell is that the perceptual object (which is elicited and maintained by many particular stimulus energies playing upon peripheral receptors) is no mere aggregate of stimuli; instead it is a system, a reconstructed inner object resulting from the systematic interactions of the various inputs. Two points may be mentioned in passing: (1) Our first impression from reading Gestalt psychology is that the perceptual object is an ephemeral thing that comes and goes depending on its stimulus supports in the environment; thus, although it may be a momentary system, it does not seem to be a relatively permanent internal system outlasting the exigencies of the environment (I will speak of Koffka's memory trace in a moment). (2) No learning seems to be involved: the perceptual object seems determined almost entirely by two factors: (i) the hereditary structure of the receptors and the brain field, (ii) the stimulus inputs which maintain the object at the moment; thus it is not a learned perceptual object, nor an "internalized" perceptual object. When particular stimulus energies are effective through particular receptors on a particular type of brain field, this is the type of organization that results.

We turn next to the perceptual field. This concept is best known from the works of Lewin. It represents, to the casual observer, a first order expansion of the perceptual object of the original Gestaltists; it treats the whole moment of experience as a single system.[15] In one sense, the perceptual field is a time-slice of experience treated as a system; it seems not to have a temporal dimension. In another sense, this cannot be entirely true, for the dynamics of the Lewin field must certainly take place in time. But these dynamics are all very short-run dynamics. They all seem to occur within a given psychological field; unless the person leaves the field. What Lewin does is to move from the object systems of classical Gestalt theory up one system level to a more inclusive system which includes perceptual objects as subsystems. There is no doubt that Lewin was moving toward the type of psychology we would like but there are certain limitations to his analyses. One senses a certain discontinuity of his perceptual fields as we move from one moment of experience to the next. When one "leaves the field" it is not quite clear what happens to the field. We would say that the subject cannot genuinely leave the field, for the field is inside of experience, not outside in the environment. So one does not really leave the field; instead, in our terms, the field goes into a latent phase. But Lewin does not conceive a particular field as organized into a temporal system of successive phases, and thus he ignores not the temporal dimension but rather the organization of experience along the temporal dimension. This, I would guess, derives from his failure to recognize clearly the distinction between structure and process. The field is a process, and the only structure taken into account is the momentary structure of that process. On a metaphysical plane, this may be a fine position; but for a detailed analysis of the personality as a system, one must take seriously the distinction between a structure which lasts, and different kinds of process that occur within the structure. From a superficial reading of Lewin, this distinction seems to be overlooked. We would suggest that a par-

[15] Lewin's life space is not just perceptual or a matter of experience. The fact that it is an inferential construct distinguishes him methodologically from the phenomenologists.

ticular field can, and regularly does, pass through several phases: it can be merely a latent structure, or at another time it may be the seat of a thought process, or at still another time, the seat of a perceptual process, and so forth.

Koffka in his discussion of memory traces begins to take seriously this problem of the distinction between structure and process. Taking his point of departure directly from the perceptual object of the original Gestalt movement, Koffka suggests that each momentary perceptual object leaves some relatively lasting structural trace. A new process (a new momentary perceptual object) can later recur in this old trace. With the trace, Koffka would like to overcome both the ephemeral character of the perceptual object, and the insensitivity of early Gestalt theory to problems of learning. The trouble here seems to be that Koffka fails to realize what a lively and active thing a "trace" can be, particularly when it is what we would call an "internalized social object." We tend to imagine Koffka's trace as the residuum of a stimulus input, but it seems difficult to fit into this category the rewarding and punishing internal surrogate for an absent rewarding and punishing human being. What is missing, I do not quite know; certainly Koffka's trace is dynamic in a certain sense of that term. Changes occur during its "latent phase." But what are these changes? Chiefly progress toward stability, symmetry, praegnance. But also it is affected by other processes, and these may cause different changes. But the changes are all the type of thing that would occur to an oil spot on water. In the absence of constraint, it takes on ever better symmetry. Given constraints, it forms the best form it can in spite of them. But it does not "fight back." It does not have internal sources of motive force, distributed differentially among its parts. Somehow, psychology has to find a way for getting motivation not only into the personality of the actor being analyzed, but also of getting different and autonomous motives into the internalized objects of that actor. The child has a concept of mother. This concept persists over time. It can be latent, or conceived, or perceived. But in any of these states, it is a concept of a loving, punishing, rewarding, wanting, mother. The internalized mother is not at the mercy of the rest of ego's wants. Rather, the internalized mother can fight back, even though the real mother is not present (even though the internalized mother is in a state of thought or latency rather than perception). The dynamics of Koffka's traces are not sufficiently dynamic for us; they are pretty sleepy dynamics. But certainly the memory trace is another step in the right direction, for it represents an attempt to give persistence over time to the psychological subsystems which we have called the "internalized objects."

On the other hand, both Koffka and Lewin have ignored a very important basis of systematization. This is the antecedent-successor relation. It may obtain between internalized stimuli, or internalized objects, or internalized fields. It is crucial with respect to the organization of objects or stimuli into motivational subsystems.

Both classical Gestalt and Lewin theory tend to emphasize what might be called spatial or perceptual relations, and also relations of similarity: chiefly the type of relations that can be described by topology or by drawings on a sheet of paper. Barriers stand between the self and an object. Two objects may have a relation of proximity, and they may form a good or bad Gestalt. Similar stimuli may be grouped to form a single Gestalt, and so forth. But the most important organizations of stimuli for the pursuit of goals, for the instrumental use of means objects, are organizations based on the antecedent-successor relation between stimuli or objects. Internal objects are organized into cause-effect chains along precisely this antecedent successor dimension; and this dimension finds no genuine acceptance in either classical or Lewin Gestalt theory. But certainly the memory trace is another step in the right direction, for it gives persistence over time to the object-system.

Tolman, however, with his sign-gestalts makes this relation the dominant principle of organization. Thus, Tolman provides the basis for the integration of stimuli into a different and cross-cutting type of subsystem. Classical Gestalt provides for the integration of stimuli which appear together into some sort of object system. Lewin provides the basis for the integration of objects into higher order field-systems. Tolman, however, provides for a different dimension of organization. Stimuli, or objects or fields may, by Tolman's theory be integrated into sign-significance systems. We may think of them as cause-effect systems; or means-end systems. A means-end system is a structural organization of internalized objects or stimuli or fields into what Tolman calls sign-gestalts, or sign-significates. One internalized object is the sign of the relation; another is the significate of the relation. The two are related by what Tolman calls a direction-distance. We may think of the one internalized object as the cause, the other as the effect, and the direction-distance is the operation which must be performed on the cause to produce the effect. Or we may say the one is the means, the other the end, and the direction-distance is that which you have to do with the means to get the end. It is important to note that this type of sequential organization transcends the organization of stimuli into objects or fields; for a given sequential organization (a given sign-gestalt) may bind together some one aspect of a given object with some other aspect of the same object, or it may bind together one object with another object, or one field with another field, or it may relate one aspect of one object with one aspect of a different object. In a sense, then, this sequential organization which Tolman takes account of, is a different mode of organizing into systems the same elements which by different principles or dimensions of organization are organized into different kinds of systems, e.g., objects or fields.

From the present point of view, Tolman thus makes a distinct advance beyond the position of the more "orthodox" behaviorist school centering on the name of Hull. Essentially, as I see it, this consists in treating a given action or behavior sequence as a system which is organized over time. This is what is meant by, or at last implied in the purposive character of behavior. It means essentially that each minimal time-sector of the process is no longer treated as immediately and directly dependent on a process, externally (stimulus) or internally (drive) operating from "outside" the action system "pushing" it through a series of changes of state, but provision is made for an internal boundary-maintaining integration of the system so that energy invested in the goal may "flow back" through a feed-back process to the instrumental acts necessary to attain the goal.

In summary of these various systematic advances made after Gestalt psychology, we may say the following. First, the original Gestalt contention that the object of perception is no mere mosaic of its component stimuli but is rather a systematic integration of them was a decisive first step in the direction we would like to see. Second, the analysis of the psychological field as a super-ordinate system including objects as subsystems was an important step in the direction of understanding what is here called the macroscopic-microscopic dimension; it is an explicit recognition that action phenomena are always made up of systems within systems. Third, Koffka's production of the memory trace was an explicit attempt to resolve the problem of getting a relatively lasting internalized object; and this, we believe, is absolutely essential to progress. Finally, Tolman's sign-gestalt brought the antecedent-successor relation between internalized objects explicitly into the study of psychological systems as a means for organizing internalized objects or stimuli along a motivation-gratification dimension.

Last, but not least, I should like to discuss a few considerations about psychoanalytic theory. Here for the first time in modern psychology, I think it is fair to say, there has appeared at least an approach to a theory of the human personality as a system, in both its cross-sectional and its temporal aspects of extension, with both cognitive and motivational emphases and couched in terms of the action frame of reference.

The last assertion may seem questionable to many readers. Doubts about it may stem from two main sources. In the first place Freud himself never fully resolved the question of the relations of his conceptual scheme to biological theory. In line with the intellectual climate of his day, particularly in the medical world, he held the view that psychology would ultimately prove reducible to biological or even biochemical terms. But the most important single fact about Freud, perhaps, was his refusal to attempt to solve psychological problems by extrapolating the biological knowledge of his time; his insistence on the direct clinical study of the human personality. In so doing he evolved a conception of "instinct" (really a mistranslation of the German word *Trieb*) which was altogether different from that of McDougall.

The second reason for the difficulty lies in the circumstances of the reception of Freud into the English-speaking world. This came, on a large scale, just as the behaviorist reaction against the McDougall type of instinct theory was setting in, and Freud was very generally identified with this "old-fashioned" type of theory.[16] There was just enough plausibility in this interpretation to help prevent any serious examination of it. It was further reinforced in behaviorist circles by their dogmatic refusal to consider the treatment of "subjective" data as scientifically admissible, and of course by the elementaristic bias which has been discussed above.

Freud's empirical concern with phenomena of psychopathology seems for two reasons to have had an important influence in direct-

---

[16] This was the interpretation of Freud which I—first encountering him in the early twenties—took for granted for a long time, and did not succeed in overcoming for more than a decade after my own sociological thinking had become much farther developed.

ing him "on the track," from the present point of view. First was the fact that clinical responsibility forced concern with the personality as a whole in a sense in which, for instance, neither the experimental psychologist in animal behavior, or in perception, nor the specialist in "testing" would have such an interest. Secondly he came immediately to a conception of intrapersonal conflict which forced his attention to problems of the organization of personality, precisely the field of problems most systematically avoided by the behaviorists.

Again, particularly to behaviorists, one of Freud's most controversial concepts was that of the unconscious. Quite apart from question of its relation to the older "introspective" psychology, which seems relatively remote today, the great importance of this conception lay in its relevance to the idea of system, in that it explicitly allowed for an "iceberg" view of the personality, including the ideas both that some parts are relatively dissociated from the more visible ones, and that there are phases in the development of motivational processes, and hence a possibility of "latency" without dissolution of the motivational unit. Without some such conception it would have been impossible for Freud to develop a theory of the growth of personality as a system.

With regard to "instinct" in the Freudian sense, the important point is its extreme non-specificity. Here again we have been led astray by our tendency to interpret Freud in terms of our own psychological common sense, and partly by his own unclarity. But the association with "sex" as we have tended to understand it has been an endless source of difficulty. The fact that Freud ended up with a duality of instinctual forces is far less important—although it involves serious difficulties—than the fact that he directly and definitely abandoned the attempt to derive the main foci of the structure of personality from specific instinctive, i.e., biologically hereditary foci.

It is here, perhaps, that the most important contribution of psychoanalytic theory from the present point of view is to be found. For, dominated mainly by the genetic point of view, Freud set out to construct a positive theory of the role of object-relations in the organization of personality as a system. This was an enormous step with the implications of which most of academic psychology has not yet caught up.

Freud saw very clearly the enormous importance of early experience in the family for the personality development of the child, and has given us classic formulations of various aspects of the child's relations to the parents. He also, with the concept of the superego, gained the fundamental insight that value-patterns of the culture come to be internalized in the personality, and that this occurs through relations to social objects, first in the family. This, by striking contrast with all the other movements reviewed, was a psychology which connected directly with the main theoretical interests of sociology (the phenomenon of internalization had also been understood a little earlier and independently by the sociologist Durkheim).

This of course is by no means to say that the integration of psychoanalytic theory and sociological theory does not present difficulties, some of them very formidable.[17] The sociology available in Freud's time was itself only a beginning and for understandable reasons he was not in touch with most of what there was. He never adequately analyzed the object system, of the child or the adult, as a system which, we have seen, is tantamount to saying that it is a social system. For his theory of development he did not have a technical analysis of the family as a system, or of its articulation with other subsystems of the society. He furthermore did not have an adequate conception of the balance between constant elements in all kinship systems and those aspects which are variant from one system to another.

On the level of the still more macroscopic analysis of social systems, as in his studies of group psychology, of religion, and the like, Freud, and many of his followers, have tended to develop a kind of "projective" sociology in which the psychoanalytic theory of personality is made to serve double duty as both a personality theory and a social system theory. But with better sociological theory available, with great accumulation of empirical knowledge, and with competence *on both sides* in a few of the same people, these difficulties are proving soluble. The difficulties which remain are secondary as compared with the fundamental significance for sociology of the reorientation which psychoanalytic theory has given to psychology. In its essential features this reorientation is not completely confined to psychoanalysis, or even movements strongly influenced by it, but its influence in this direction is so much the most prominent, that it has seemed fair to concentrate on it for present purposes.

[17] The "psychological" movements laying greatest stress on social interaction as such are of course those associated with the names of Harry Stack Sullivan and J. L. Moreno. We cannot take space to go into them here.

IV

Only a few brief words need to be said in conclusion. This paper has concentrated on a few aspects of the larger picture of the relations of sociological and psychological theory, particularly those directly relevant to the systematic character of the two conceptual schemes in relation to each other. This concentration of interest has introduced a certain selectivity of emphasis, in the sense that it fails to do justice to the possibilities of collaboration between psychology and sociology where the type of theory on one side or both fails to meet the specifications which have been laid down here. I have in no way meant to imply that nothing from the work of instinct theorists, of behaviorists, or of Gestaltists, is of any use to sociologists. On the contrary, much of it has proved to be extremely useful. Nor does it imply that everything in psychoanalytic theory is sound and useful to the sociologist. But this fact does not dispose of the problem on the systematic levels on which they have been considered in this paper.

Perhaps the most crucial problem is that of the macroscopic-microscopic range and its implications. Our contention here is that the subject matter of the sciences of action is human behavior (with some interest in the subhuman of course), and that empirically this is subject to study on *any* level on which it is found. Possible cyclical patterns in the history of great civilizations each lasting for centuries are, subject to the availability of empirical evidence and of techniques of getting and validating it, just as legitimate objects of scientific study of behavior as are the cycles of wakefulness and sleep of a single individual, or the maze behavior of a rat.

Beyond this, we have now accumulated ample evidence for the view that the same basic conceptual scheme is applicable through an indefinite range in this respect. Of course many different "adjustments" have to be made in the use of theory at different levels, but these are of the same order as those used in physical science; they involve different empirical operations, and different formulations, but not a fundamentally different theoretical framework. There is, to be sure, one fundamental "break" which occurs in the "asymmetry" between social system and personality, but it has been a crucial thesis of this paper that this does not involve a fundamentally different frame of reference or set of categories, though many of the generalizations about systems will of course be different.

It follows further, that no one "level" in this range, including the "shift" between social interaction and personality system, has any claim to ontological priority. The "theory of behavior" in the Hullian sense has no priority, except that claimed on grounds of better study and greater precision, over the "theory of history" in a sense of analysis of social processes taking thousands of years. Data from both are equally valid as contributing to a general theory of action.

This is the focus of the crucial problem for the relations of sociology and psychology. As we have learned from Whitehead (the "fallacy of misplaced concreteness") and from Morris Cohen (the fallacy of "reification") this tendency to illegitimate reification is an endemic "disease" of our Western intellectual culture. The movements in psychology just reviewed have tended to be guilty of this on three different levels: (1) reifying the organism, which by virtue of its genetic constitution is alleged to provide the "real" basis for the structure of behavior systems; (2) reifying the "real unit of behavior" which may be either the S-R sequence of the behaviorists, or say the momentary perceptual "gestalt." This then is regarded as the one key to the understanding of all behavior or; finally, (3) reifying the individual, the personality in a more or less clearly defined "action" sense. On knowledge of him, independently of his social relationships, current or previous, is alleged to depend any genuine understanding of how individuals, when put together in societies, will behave.

I submit that *any one* of these reifications or *any combination* of them constitutes a barrier to the progress of social science. All of them—and the corresponding reification of "society"—are barriers to the recognition of the *relativity* of perspective on systems which, in the physical sciences as in our own field, has proved to be the most important principle of theoretical advance in science. It is only when sociological and psychological theory—and the various system-reference levels within each—have come to regard themselves and each other as the formulators of very important *special cases* relative to a more general theory, that a higher level of theoretical maturity in our fields will have been reached. The contention that any one of these fields or "levels" provides the "foundations" on which all the others *must* build or be consigned to the hell of perpetual scientific impotence, is only a symptom of the growing pains of a very young family of scientific disciplines.

CHAPTER 5

Anthropology and Sociology

HOWARD BECKER

I

IN THE "FUNNY PAPERS" of a bygone day there appeared two characters named Alphonse and Gaston. Their mutual courtesy was mar-
velous to behold; so likewise were the predicaments in which their
mutually obstructive "After you, Alphonse," and "After you, Gaston"
perpetually involved them. Usually a hearty kick from Maud the
Mule provided the only feasible solution. The plan of this symposium
would result in a certain Alphonse-Gastonishness if the sociologists
dealing with the anthropologists, and the anthropologists with the
sociologists, and so on around the circle, attempted to bow each other
through doors and over fences too much, more especially as our
editor is not inclined toward the use of Maud the Mule's method.

Ergo, the writer of this chapter feels that the purposes of all concerned are best served by discussion that is entirely frank. It need
not therefore be tart or blunt; Alphonse and Gaston do not find their
necessary alternatives in Captain Flagg and Sergeant Quirt of "What
Price Glory?" fame. If we were like such old-fashioned professional
soldiers we might knock each other down one minute and step up to
the bar together the next, but we aren't.

Disregarding any "More's the pity" that may be muttered, let it
be announced that the main features of the present chapter are first,
an attempt to say, plainly and directly, how anthropology and

sociology, as going concerns, came to be what they are in several important countries; and second, what facts, concepts, and procedures of contemporary anthropology are or should be useful to contemporary sociology.

Both disciplines, if such they may be called, have at various times issued pronunciamentos laying claim to the entire human world and its ways, but these claims serve, in the writer's estimation, only to illustrate a recurrent variety of academic imperialism. Moreover, the fact that anthropology has "contributed" to sociology, and *vice versa*, and that in proper combination the same might be said of social psychology, does not mean that the resulting Grand Alliance has a monopoly of worthwhile scientific knowledge of *Homo sapiens* and his affairs. There are other "sciences of man," social sciences, or behavioral sciences that might well be considered along with those included in this symposium. Every shoemaker thinks there is nothing like leather, true enough, but we are or should be citizens of an intellectual commonwealth as well as specialized craftsmen.

It would be gratifying if the discussion could begin with definitions of anthropology and sociology that are everywhere accepted, but such definitions couldn't be found even if "everywhere" were changed to "everywhere in the United States." Hence talk in general terms about the "contributions" of anthropology to sociology is impossible to carry on without falsification or, at the very least, confusion. "When, where, and how does who contribute what to whom?" is a pointed kind of question that may lessen the confusion even though it increases the number of facts to which reference must be made. After such a question has been answered a number of times in a number of contexts, anthropology and sociology should, in a way, define themselves. If they don't, a good stock of answers may nevertheless make it easier for others eventually to present definitions having fair chances of acceptance.

To amass that stock of answers it seems necessary, to the writer at any rate, to do first what he has already said would be done first; namely, to show how anthropology and sociology, as going concerns, came to be what they are in several important countries. Then the matter of contributions can be dealt with much more easily.

The choice of the countries in terms of importance here depends, naturally enough, on importance judged with regard to the furnishing of significant leadership in anthropology and sociology. This would seem to mean that there is also the possibility of using the criteria of "key personalities," "schools of thought," and the like. Such possibility, however, is limited; as the writer has put it elsewhere:

. . . we shall permit national boundaries or the limits of language to define the fields surveyed . . . [for in contrast to the natural sciences, in the social sciences] the central core of any theory one cares to name is almost inseparably connected with the language in which it is formulated; even statistical generalizations provide only a partial exception to this rule. Not only this: the cultures of the different countries in which the social sciences have developed possess a number of aspects other than language which vitally affect the . . . theories that develop within them. It may some day be possible to transcend such "cultural compulsives," [but until then] . . . we must continually remind ourselves that the cultural settings in which . . . [social-scientific] theories are placed oftentimes make writers whose points of view are widely separated, but who share the same culture, much more like each other than those holding apparently similar views [but] in widely contrasting cultures.[1]

Otherwise put, it is this cultural diversity that helps to account for some of the peculiar failures of social-scientific cross-fertilization, on the one hand, and for certain kinds of inbreeding, on the other. Personalities and schools of thought should receive proper attention, of course, but rarely will they be found to transcend their cultural limitations.

The countries surveyed will be the French-speaking, the German-speaking, and the English-speaking, for they are probably, on any basis, the most important in past and present social science, and are reasonably familiar to the writer. The period covered will be from about the seventeenth century to the present, although as far as even moderately extensive reference is concerned the span from the mid-nineteenth century onward is treated as of primary importance, with the present end of the span bearing by far the greatest weight.

"Where" and "when" in our question about "contributions" thus delimited, there remain "what," "how," "who," and "to whom." These will be taken in general and oftentimes implicit senses, and answers, although sometimes detailed, will seldom if ever run through a rigid "what" *et cetera* sequence. Instead, there will be reference, as occasion requires, to major discoveries, writings, methods of investigation, and shifts in the climate of opinion; to the occupations of the

[1] Howard Becker and Harry Elmer Barnes, *Social Thought from Lore to Science*, 2nd ed., Washington, D.C.: Harren Press, 1952, Vol. 2, p. 793.

"contributors"—academicians, missionaries, reformers, museum workers, colonial administrators, and whatever; to the audiences or publics, both of non-professional and professional varieties, who received the "contributions" or by whom they were effectively passed on; and to anything else that seems relevant!

II

Social-scientific developments in the homelands of French culture have manifested, early and late, a strong interest in "natural reason" and "natural man." This was apparent in realms as diverse as *The Jesuit Relations,* on the one hand, and the writings of Rousseau, on the other, and in periods as widely separated as that of the North American explorations and that of North African expansion. Such an interest may be caricatured thus:

The French believe that when man is truly natural and truly reasonable he is, in essence, truly French, and hence that all civilization of natural and reasonable type is at bottom French civilization. The famous "rights of man" are thus simultaneously the rights of Frenchmen, and when human nature realizes its culminating potentialities it necessarily produces the Parisian.

The caricature obviously blurs and distorts and it of course exaggerates the French aspects of that theme so very common among all peoples; namely, ethnocentrism. Nevertheless it may crudely outline a somewhat special set of attitudes that, in spite of the counter-instance of a Gobineau, can be regarded as characteristic of many intellectual Frenchmen—attitudes making for tolerant, serenely confident interest in "primitives" of every sort.

The tolerance and confidence, if the characterization may be tentatively continued, derive from the fact that cultural relativism and cultural absolutism are happily reconciled in this conception of the natural. No culture, however primitive, need be rejected, for it is primitive only as an embryo is primitive; given time and opportunity, its inherent nature will become increasingly manifest in ever-closer approximation to that universal model most attractively bodied forth in Marianne. A better analogy, perhaps, is that of the snake's skin; necessary to the growth of the organism enclosed, it is naturally superseded from time to time, but can nevertheless be admired for the effective way in which it fulfills its transitory function.

These analogies, however, do not quite make the point, for they are cast in biological terms, and the attitudes being discussed were in evidence long before biology was a stylish source of metaphor. Certainly no evolutionism, latent or otherwise, was in any way intended. A better image is that of cultures as rough jewels, each one variously displaying a number of polished facets but with the bulk of the stone in greater or lesser degree uncut; French and *ipso facto* universal culture is that which, with all its facets shooting forth the light of reason, elicits in maximum degree the potentialities naturally present. Natural and artificial therefore do not stand in opposition; on the contrary, art alone can fully realize nature. The rough jewel may therefore be cherished for what its few flashing facets emphasize; only through its fitful and partial radiance can the wonder of the perfectly cut gem be fully appreciated.

Awareness of cultural relativism and devotion to cultural absolutism, then, need not have been mutually exclusive if the relativism and absolutism actually did take the peculiar forms just presented—and indeed, there is demonstrably much in the intellectual tradition of the bearers of French culture to show that they did. Take, for example, the long-sustained interest in *l'histoire des moeurs*. Long before Voltaire, long before Montesquieu, the mores of urbanity and polite society were contrasted with those of both internal and external barbarism. The barbarians were viewed as diamonds in the rough, *à la* Caractacus or Vercingetorix, and exhibition of the fascinating polish, now here, now there, effected in the manners of their descendants filled many delicately limned pages and columns of commentary. As might be expected, variability in *les moeurs* generated erotic as well as esthetic interest, but in French life these were not sharply distinguishable. Moreover, the variability also came to be of interest for its own sake in many cases, but to this purely intellectual attraction was often linked a reformist allure; moral satires such as Montesquieu's *Persian Letters* or political tracts such as Volney's *Ruins of Empires* were the result. In short, cultural relativism of several types showing considerable similarity to the tolerant, confident, absolutist-anchored kind here discussed has long been evident in European regions dominated by French culture, where even today it contributes to a relaxed attitude toward "colonial peoples" who show signs of becoming good Frenchmen.

At its best, such an attitude helps the French observer, amateur or professional, to enter into the intimacies of the people he studies more completely than is usually the case among those who tend to view nature as radically evil. Now, intimate detail, when appropriately communicated, generates sensitivity to the nuances of values and value-systems, and thereby restrains the crudity of economic, geographic, and similar determinisms. Even those using evidence collected by observers outside the orbit of French culture were apparently so alert to subtleties that they perceived—or at the very least stressed—in the evidence what the initial observers had regarded as inconsequential. To cite a relatively recent instance: the British anthropologists Baldwin Spencer and F. J. Gillen, on whose Australian field work the French sociologist Durkheim so heavily depended, were much less keenly aware of certain of the implications of their discoveries than was Durkheim. He had profited by his possession, among other things, of the widely diffused French flair for what Hugo called "the perception of the intangibles that make all the difference."

At its worst, the tolerance of the peoples of French culture for the native diamond in the rough led (and occasionally still leads) to a delight in the quaint, the exotic, and the titillatingly unfamiliar *je ne sais quoi* that, Watteau-like, made coy shepherdesses and languishing swains out of somewhat more earthy characters. The resulting damage to sound social-scientific knowledge was serious but by no means fatal.

What might have been fatal, had it long prevailed, was Rousseau's conception of nature as opposed to culture. This bears a surface similarity to the Watteau-like tolerance just noted, but at even a slight depth the similarity vanishes. Rousseau's Savoyard vicars and Émiles owe nothing to culture; their personalities unfold like flowers under the sway of immanent vital impulses. To return to the figure previously used, the glow of their native goodness is not the result of an art that shapes and smooths the essential facets; they are gems that grinding and polishing can only distort—yes, destroy. With this notion of the natural the conceptions of the *philosophes* and, indeed, of most "intellectual" representatives of French culture were at almost complete variance; Rousseau's naturalistic "enthusiasm," a French-Swiss equivalent of Methodism with pantheistic trimmings, was viewed with distrust and even disdain. Only his political doctrines, particularly those bearing on the general will, met with thoroughgoing acceptance, and they had no necessary connection with his nature *versus* culture credo.

During the early part of the same general period during which the interest in natural man and natural reason was strongly manifest, the quarrel as to whether the Ancients were greater than the Moderns, or *vice versa*, had got well under way. The controversy itself is of little importance to us here, but among the views of the many participants those of Fontenelle, Perrault, and Turgot are of interest because of their agreement on points such as the continuity of "human nature" from earliest historical times to their own day, the biological unity of mankind, the importance of the accumulation of culture, and the clash of cultures as the principal source of growth of knowledge and power.[2] There is little direct testimony, by way of citation or quotation, that they influenced the social scientists of later generations, but their opinions strongly pervaded French literature of widely-read varieties, and it seems unlikely that the parallel passages to be found in the writings of the social scientists were of independent origin. However this may be, a relatively insignificant amount of biological determinism, to say nothing of racism, and a correspondingly heavy emphasis on culture, have long been characteristic of the French "intellectual" tradition, and anthropology and sociology certainly draw on that tradition.

The tendency to pay great heed to culture is clearly apparent in the work of even the physical anthropologists. When the Paleolithic and Neolithic sites in which France abounds began, during the nineteenth century, to receive increased attention, "cave man" was studied quite as much for his implements as for his physique. Boucher de Perthes, for example, helped to launch the eolith controversy about the most primitive culture of *Homo faber*. Were the "dawn-stones" nothing but brook-battered flints untouched by human hands, or were they really crude wedges, pounders, and choppers?

Questions about material culture such as these followed, naturally enough, on the discovery of the objects in question. Almost a half-century earlier, however, and long before the discovery of the Indo-European language family, questions about man's speech as the core of his non-material culture were being asked and confidently answered by the Vicomte DeBonald. True, this ultra-conservative used reason only to refute reason, and his answers were essentially anti-scientific, but he did insistently call attention to the part played by language in the patterning of man's conduct. He and other traditionalists such as DeMaistre passed many of their essential conceptions on to Saint-Simon and Comte, through them to French anthropology and sociology generally, and through these to the same and related sciences everywhere else. In many respects different from or

[2] *Ibid.*, Vol. 1, pp. 461–474.

opposed to the ideas thus far discussed, they too form part of the French tradition, hence it seems advisable to present a summary of their salient features:

> . . . society rests on a moral *consensus;* . . . the individual has no rights, but duties; . . . the family rather than the individual is the true social unit; . . . society is a reality above and beyond its constituent individuals; . . . the reestablishment of social stability rests upon the disciplining of the individual and a restoration of the respect for authority; . . . religion is an essential and indispensable instrument of social control. . . .
>
> [In contemporary French social science many of these ideas] . . . reappear in a new and somewhat more scientific guise, but without any fundamental change in nature. Society is a reality *sui generis;* the individual is merely a partial expression of this larger reality and is powerless to effect any social changes on his own account; social institutions and traditions cannot be given an individualistic explanation since they are "social things"; the central social problem is that of social control, which is to be achieved solely by securing the individual's submission to the moral rules of the group externally imposed; society is in essence something external to and constraining the individual; the liberation of the individual from the constraining influences of the group is a source of social danger and personal disorganization leading even to suicide; the surest guarantee of social integration and stability lies in a body of dogma which is generally accepted; and religion is essential in this connection, for it has the basic function of revivifying group sentiment and reaffirming traditional standards and values.[3]

What is here strikingly similar, in essence, to earlier French formulations—excepting Rousseau's—is the minimizing of the merely biological and the stress on the importance of culture. It must of course be granted that culture, as implicitly discussed by the traditionalists, is no ballet of bloodless categories, but culture in the Tylorian sense, namely, "that complex whole . . . [of] capabilities and habits acquired by man as a member of society." Family, social institutions, social control, and the like are cultural phenomena, but also and inseparably, phenomena of society. Witnessing the overthrow of the *ancien régime,* suffering in the turmoil of the revolution, living in hardship as *émigrés,* and experiencing the vicissitudes of the restoration, the traditionalists were fully and even excessively made aware of the fact that a culture necessarily has carriers. Although linguistic reasons are probably of greatest significance, with antagonism to the Germanic use of *Kultur* also playing an important part, it is entirely possible that the contemporary French unwilling-

---

[3] *Ibid.,* Vol. 1, pp. 498–499.

ness to distinguish between society and culture, and the use of *civilisation* as a partial equivalent of the latter,[4] derive in some degree from the transmitted convictions, right or wrong, of the traditionalists.

But if the traditionalists succeeded in effectively transmitting any of the ideas to which reference has been made, their success was not readily apparent to the outside world, and especially not to the outsiders innocent of social science, in the period from 1871 to the beginning of World War I. Indeed, these are minimal limits; almost the entire nineteenth century and the first quarter, at least, of the twentieth can be viewed as characterized by the steady advance of secularization (using this term in a special but warranted sense). "Folk-prescribed" societies such as those scattered, in varying density, throughout Brittany, Normandy, remoter Gascony, the Auvergne, Provence, and Alsace dwindled in scope and power of sacred control; the Third Republic, as the political framework of a "principled" society, increased its secularizing influence in spite of clerical–anti-clerical struggles and the near-catastrophe of *l'affaire Dreyfus;* rapid loss of generally accepted basic values led to widespread normlessness and disorganization.[5] Along with this, however, went the revival of old or the development of new sacred controls and would-be controls, particularly at ideological levels involving one or another kind of *mystique.* The extreme nationalism of a Goebbels-like Paul Déroulède or of a Hitler-like General Boulanger, hardly surpassed even by the devotional intensities at Napoleon's tomb, provides striking instances. So also do the religious assaults on *laïcisme* in the public vs. parochial school struggle, and their counterparts, *Grand Orient* Masonic and similar "parliamentary conspiracies." Still other examples are the "philo-Semitism" and anti-Semitism of the Dreyfus case, and the persisting antipathies of "proletarian" and "bourgeois" as demonstrated in the Commune, its Père Lachaise aftermath, and ensuing May-Day exhortations to revenge.

---

[4] See A. L. Kroeber et al., *Culture: A Critical Review of Concepts and Definitions,* Cambridge, Mass.: Peabody Museum, 1952, XLVII, 1 of Papers, pp. 11–12, 37–38. In addition, the remarkable study by Joseph Niedermann should be consulted: *Kultur: Werden und Wandlungen des Begriffes und seiner Ersatzbegriffe von Cicero bis Herder,* Florence: Biblioteca dell' "Archivum Romanicum," Serie I, Vol. 28, "Bibliopolis," Libraria Antiquaria Editrice, XIX, 1941.

[5] The concept of secularization has been developed in some detail in my *Through Values to Social Interpretation,* Durham, N.C.: Duke University Press, 1950, Chs. 1 and 5. A good deal of what is said in the above paragraph is not fully intelligible to those unfamiliar with this analysis. Yet, brevity is necessary.

The result was chaos; the continual colliding of value-systems led to the further crumbling of the old and the splintering of the new. Many social scientists who cherished loyalties to the supreme values of French culture as manifested in France as a nation, but who were not necessarily nationalists in the chauvinistic sense, developed a sense of mission as regards *la patrie*.

This patriotic mission was conceived to be the restoration or creation of national solidarity, and under its spell many of the doctrines of the traditionalists were revived in modified but recognizable form. Unfortunately for the exponents of *solidarisme*, the doctrines that they hoped would profoundly impress a broad public left only shallow traces on other than the academic intellectuals, a small sprinkling of right-wing politician-publicists such as Charles Maurras and Maurice Barrès, and, paradoxically enough, a few advocates of "industrial democracy" counting among their number Célestin Bouglé. Far more effective in producing solidarity in *la patrie*, for a time at least, were the pressures eventually to be exerted by World War I.

Secularization verging on normlessness was by no means stopped, however, although its rate was slowed somewhat; a seemingly inevitable progression led from the financial-parliamentary tangles of the nineteen-twenties to the Stavisky affair of the 'thirties, and from there to Blum left-wing riots, Abetz right-wing seduction, feckless Maginot *Sitzkrieg*, Pétainist military collapse, and squalid Vichy collaboration of the 'forties. Whether the Fourth Republic can develop a *mystique* that lends popular durability to its eminently logical principles remains, in the 'fifties, still to be seen.

This is a digression on the theme of secularization today, and attention should again be directed to the early part of the century. Among the social scientists who developed the sense of mission resulting in *solidarisme* was Émile Durkheim. Endowed with logical acumen, capacity for hard work, altogether unusual ability to concentrate on a few fundamental themes, and systematic rigor, he had in addition the power to gather about him a "school" of social scientists who either furnished creative additions to his basic formulations or at the very least made correlative reference to whatever parallels appeared between their results and his. He is clearly a sociologist in the Comtean line of descent, and as such will be dealt with elsewhere in this symposium.

What is of importance for us here is not Durkheim's direct influence on the sociologists or anthropologists of his own time, but rather his "hop-skip-and-jump" effect as evident today. Influenced by historical functionalists such as Fustel de Coulanges, and receiving from other forerunners the conception of society as an organism, with institutions fulfilling the functions of the various bodily parts, he transferred that conception to the mental level. Individual minds share in a *conscience collective* (which must be rendered as "collective consciousness," *not* "collective conscience"). This is a clumsy and misleading way of talking about what has since come to be called symbolic interactionism, as developed by Wilhelm Wundt, George Herbert Mead, Ernst Cassirer, Suzanne Langer, and others. For all his clumsiness, though, Durkheim avoids most of the disastrous consequences of "group mind" notions of the sort expounded by William MacDougall, and so do nearly all his followers. In fact, Glotz the historian, Schuhl the Hellenist, Gernet the jurist, Lévy-Bruhl the philosopher-ethnologist, and Granet the Sinologist—each one in his way a loyal Durkheimian—deal with symbolic systems, and particularly with language, as entirely incorporating the "collective consciousnesses" of their respective societies. Hence, even for the unkindest interpreters of Glotz et al., there can remain no supersensible residue, no brooding omnipresence in the sky, to form a "group mind." The symbolic system of any given society, which system is the key component of its culture, is its *conscience collective*.

Side by side with Durkheim's stress on this aspect of culture goes a not altogether congruent emphasis on "morphological" factors in society, i.e., population movement, distribution, density, etc., and the group structures that are presumably the consequence of such population conditions. Here he introduces his conceptions of mechanical and organic solidarity and, in developing the integrally related conception of the division of labor, arrives at a statement of structure and function that parallels his development of similar themes in his discussions bearing on *conscience collective*.[6]

The significance of this in the present context lies in the fact that Radcliffe-Brown, the British anthropologist, was strongly impressed by Durkheim's ideas about structure and function, and through Radcliffe-Brown one of the most articulate types of anthropological functionalism developed. (This is by no means to deny that other sources, influencing Malinowski and those sharing similar views, played important parts in its growth; moreover, there is no denial of the fact that there are several versions of functionalism.) The contemporary vogue of structural-functional sociology in the United

6 Becker and Barnes, *op. cit.*, Vol. 2, pp. 829–839.

States in turn owes a great deal to anthropological functionalism. This arises from the fact that, although Spencer, Sumner and other sociologists long ago made much of structure and function, the ready acceptance of similar ideas today, particularly where those having to do with function are concerned, is in considerable measure the result of their recent popularization by anthropologists such as Margaret Mead, Gregory Bateson, Ruth Benedict, W. Lloyd Warner, and Robert Redfield.

And this is what was meant above by Durkheim's "hop-skip-and-jump" effect; he made an impact on anthropologists that they have transmitted to sociologists. Let it be reiterated, however, that today's anthropological functionalism incorporates a good many features that are not traceable to him, and that those so traceable have inevitably undergone many changes in process of being passed on—"inevitably," for neither he nor anyone else can be fully understood out of context. *La patrie* and *solidarisme* are among the symbols essential to that French *conscience collective* in which Durkheim participated and to which he contributed so much.[7]

Difficulties of definition have thus far caused us little concern; they have been casually avoided or planfully dodged. The point has now been reached, however, when a few comments about the meaning of terms may not be amiss.

In the French-speaking world, anthropology ordinarily signifies only what we would call physical anthropology, or even anthropometry. In the foregoing pages, French usage would have required either ethnography or ethnology, or both, where anthropology has

---

[7] If space permitted, considerable attention would be given to Lévy-Bruhl, whose theories about "primitive mentality" have been sadly misunderstood (see the recent exchange of letters in the *British Journal of Sociology*, Vol. III, 2 July, 1952, pp. 117 *et seq.*, to Henri Hubert, Marcel Mauss, and many other Durkheimians, to Claude Levi-Strauss, and to anti-Durkheimians of the sort represented by Gaston Richard). But see Becker and Barnes, *op. cit.*, *passim*, especially Vol. 2, and Georges Gurvitch and W. E. Moore, *Twentieth Century Sociology*, New York: Philosophical Library, 1945, article by Claude Lévi-Strauss.

Incidentally, there has been shameful neglect—"space" is a poor excuse—of important French-Swiss and Belgian figures in all of the foregoing. The Belgian anthropologist Van Gennep, for example, has made a very important contribution to sociology with his conception of *rites de passage*, probably best translated as "ceremonies of transition." Instance the use made of Van Gennep's ideas by J. H. S. Bossard and Elinor H. Boll, *Ritual in Family Living*, Philadelphia: University of Pennsylvania Press, 1950; the sociological application has been direct and effective. American anthropologists, oddly enough, have made little systematic use of Van Gennep's *rites*. An important exception to this statement is Eliot D. Chapple and Carleton Coon's *Principles of Anthropology*, New York: Holt, 1942, but it is among the very few.

occurred without the adjective "physical." Anthropology, etymologi-
cally the science of man, applies in French merely to the science of
man as biological specimen. Ethnography, in its literal sense the
relatively particularized description of a people or peoples, is definitely separated from direct biological considerations, as is also
ethnology, which by derivation and custom has come to mean the
generalizing science of peoples.

Throughout all of this the social scientists of the French-speaking
countries (indeed Continentals generally) faithfully follow the original Greek denotations and connotations of *anthropos* and *ethnos*,
graphai and *logos*. Only recently has the American trend toward a
more inclusive use of anthropology as designating all the sciences
(or even *the* science) dealing with things human begun to find any
echo in Europe: Penniman, Lévi-Strauss, and Mühlmann—English,
French, and German respectively—provide relevant passages in their
current writings.

American anthropological practice, however, still differs considerably from the implications or explicit claims of American anthropological terminology. In spite of emphasis on *the* science of
man in what has been called, perhaps unkindly, "the new anthropology,"[8] the anthropological neophyte must in nearly all cases
establish himself professionally by studying and, if possible, doing
field work on some "primitive" or "simple" people and their culture.
"Primitive" and "simple" have recently come to include rural American villages, folk societies in Yucatan, German small towns, Scottish
Border parishes, and California "industrial farms," but these are
ordinarily viewed as "wholes" lending themselves to analysis by the
same procedures as those usable for small Australian tribes or
Zuñi pueblos. With the moot epistemological and other issues arising
from the sometimes explicit holism and configurationism involved
there need be no concern at this point; for present purposes it is
enough to assert that in at least the early stages of their careers
American anthropologists demonstrably utilize assumptions and
procedures differing in essential respects from those characteristic of
American sociologists.

In the regions under the sway of French culture there is even
today no such readily noticeable difference, nor was there in Durk-

[8] Robert A. Endleman, "The New Anthropology and Its Ambitions," *Commentary*, 8, Oct., 1949, pp. 284–291. Endleman is probably tilting, consciously
or unconsciously, at Clyde Kluckhohn's *Mirror for Man: the Relation of Anthropology to Modern Life*, New York: McGraw-Hill, 1949.

heim's time. Indeed, ethnographer, ethnologist, and sociologist were occasionally combined in the same "triple-threat man," as in the case of René Maunier, who did excellent field work on the Kabyles of North Africa, generalized about "colonial peoples," and was one of the most competent urban sociologists of the early part of the present century.[9] Social scientists representing the mere "double-threat" combination of ethnologist and sociologist are so numerous that any brief listing is highly arbitrary, but there may be mentioned the well-known names of Georges Davy, Henri Hubert, and the recently deceased Marcel Mauss.[10]

In fine, where the French-speaking social scientists are concerned it is still hard to say just when, where, and how who contributes what to whom; they are the interdisciplinarians *par excellence*.[11]

III

Without going deeply into the tangled problem of exactly when *cultura* and its related Latin forms underwent direct or indirect transmutation into *Cultur* or *Kultur* in German usage, it can be said without qualification that between 1663 and 1675 the famous German exponent of natural law theory, Samuel Pufendorf (1632–1694), offered definitions of *cultura* that, taken together, represent the full equivalent of Tylor's definition of two centuries later![12] Leibniz, in his German writings dating from about 1700, uses *Cultur*, but only

[9] Becker and Barnes, *op. cit.*, Vol. 2, pp. 859–860.

[10] *Ibid.*, pp. 839–842. See also the recent collection of Mauss's essays, *Sociologie et Anthropologie*, Paris: Presses universitaires, 1950.

[11] This is made especially clear in Claude Lévi-Strauss's introduction to *ibid.*, and also in the section entitled "Place de la sociologie dans l'anthropologie," pp. 285–290.

[12] So sweeping a statement needs full documentation. Fortunately, it is provided in the remarkable treatise by Joseph Niedermann, 1941, already mentioned on p. 110; Pufendorf's writings, even to the quite obscure letters, are quoted in detail. Moreover, the detail is sufficient to show that Pufendorf's final achievement of an array of definitions resulting in the equivalent of Tylor's was accompanied by an oftentimes quite arbitrary treatment of traditionally correct Latin usage. Indeed, one writer says that "only a terrible Latinist like Pufendorf" could so alter the content of the concept that it would take on in the seventeenth century the sense it is thought to have aquired only in the nineteenth, when Latin as the language of scholarship had been almost entirely superseded by the various European vernaculars. (E. Hirsch, "Der Kulturbegriff: eine Lesefrucht," *Deutsche Vierteljahrschrift für Literaturwissenschaft und Geistesgeschichte*, Vol. III, 3rd year, 1925, p. 400.) For comment on Hirsch's treatment see H. Günter, *Deutsche Kultur in ihrer Entwicklung*, Leipzig, 1932, p. 267, as well as Niedermann, *op. cit.*, pp. 153, 184.

in the sense of personal cultivation, as, e.g., a cultured or cultivated man.[13] The influence could have been directly through Latin or indirectly through French, for in the latter language, *culture* in this meaning appears as early as 1558,[14] and Leibniz's French-writing proclivities are well known. Evidence for the emergence of *culture* in Germany in the modern and more relevant sense is provided by another French-writing scholar, F. V. Toussaint, who in his Berlin Academy essay of 1765 discussed *"Des inductions qu'on peut tirer du langage d'une nation par rapport à sa culture et à ses moeurs."*[15] The concept of culture therein set forth also comprises certain aspects of cultivation, but Niedermann is nevertheless justified in saying that Toussaint provided "a fully inclusive, modern concept of culture."[16]

But whether the source of German usage was directly Latin or indirectly French, that usage was firmly established by 1764, for this is the latest possible date of the formulation of Herder's *Versuch einer Geschichte der lyrischen Dichtkunst,*[17] and in this essay both *Cultur* and *Kultur* appear frequently—Herder was not consistent in his spelling—with every indication of complete absorption into commonly-understood German. Like Toussaint's, Herder's concept included a considerable component of cultivation, but at many points he used *Bildung* to refer to this, leaving *Kultur* to indicate a somewhat less personal and more objective aspect of the non-material achievements of mankind.

All in all, however, Pufendorf's treatment of *cultura,* more than a century earlier, brings him appreciably nearer to modern definitions of culture than does Herder's handling of *Kultur,* close though the latter undeniably comes.

But neither Pufendorf nor Herder use *Kultur*en; in fact, as late as 1821 a German lexicon carries the comment "without plural."[18] Not until the day of the great naturalist and traveler, Alexander von Humboldt (of whom more will be said later), was attention widely and successfully called to the fact that mankind has many culture*s.*

[13] Niedermann, *op. cit.,* p. 178.
[14] M. P. E. Littré, *Dictionnaire de la langue française,* I, Paris, 1863.
[15] In *Mémoires de l'Academie Royale des Sciences et Belles-Lettres,* Année 1765, Berlin, 1767, pp. 495–505.
[16] Niedermann, *op. cit.,* p. 213.
[17] *Ibid.,* pp. 214–218.
[18] J. C. Harnisch, *Handwörterbuch der deutschen Sprache,* Leipzig, 1821.

It was perhaps as a result of his amazingly popular lectures of 1827–1828 that this occurred, for these formed "the cartoon of the great fresco of the *Kosmos*,"[19] the *magnum opus* in which the plurality of cultures reached full expression, but which did not begin to be published until 1845. However this may be, after the date noted *Kulturen* began to come into general use—although Klemm, often mentioned along with Waitz as among the Germans from whom Tylor borrowed his usage, did not use the plural even in his *Allgemeine Culturwissenschaft* of 1854–1855.

Much of the trouble in the tracing of *Kultur* in German arises from the fact that it is a loan-word; questions about the use of the native German *Volk* and its derivatives are by contrast answered with refreshing ease.[20] *Kultur* and *Volk* are here mentioned together because the latter often served as a near-synonym of the former, particularly in the writings of Herder and those influenced by him. It may be that one reason for the slowness with which *Kultur* acquired a plural form is the fact that *Völker*, the plural of *Volk*, provided a convenient substitute. The sense of *Volk* is always that of "people" or "a people," herein paralleling the Greek *ethnos*. Accordingly, *Volkskunde* is an equivalent of ethnography (although often with a meaning close to that of folklore), and *Völkerkunde* matches up with ethnology quite well in spite of the fact that -*kunde* has the concrete descriptive sense of *graphē* or *graphai* rather than the abstract generalizing sense of *logos*.[21] The match in part results from the comparative and hence inevitably generalizing implications of the plural in *Völkerkunde*; "description of peoples" rather than of "a people" necessarily goes beyond particularization in the direction of ethnology.

Moreover, it is readily demonstrable that the history of culture and the "-ology" of peoples were frequently paired in the German-speaking world from at least the third quarter of the eighteenth century onward. The pairing was especially evident in Herder's epoch-making treatise on the theory of history;[22] culture and its carriers

[19] Karl Bruhns et al., *Alexander von Humboldt, eine wissenschaftliche Biographie*, 3 vols. with bibl., 1872, Eng. trans. by Lassell, 1873, as quoted in *Encyc. Brit.*, 14th ed., Vol. 11, p. 878.

[20] Kroeber et al., *op. cit.*, Appendix A.

[21] However, Meyer in *ibid.* states that that *Völkerkunde* is synonymous with ethnography. He does not refer to *Volkskunde*.

[22] J. G. von Herder, *Ideen zur Philosophie der Geschichte der Menschheit* (first published, 1784–1791), in *Saemmtliche Werke*, Berlin: Suphan, 1887.

were presented together. To be sure, this work of Herder's represented far more than sustained attention to both *Kultur* and *Völker;* his "philosophy" was essentially an effort to state the recurrent regularities in their interrelations, with special reference to regularities in the succession of changes. What English-speaking anthropologists have recently come to call the diachronic, as opposed to the synchronic, was clearly of great interest to Herder; he was quite as much concerned with sequential as with simultaneous relations.

Otherwise put, Herder concerned himself both with the general ways in which culture grows and changes and with the particular ways in which the various carriers of culture fulfill their tasks—with the development of the culture of mankind and with the unique configurations of peoples concretely epitomizing various aspects of that culture.

This late-eighteenth and early-nineteenth century German interest in the development of culture can be thought of as in some respects like the French cultural absolutism already discussed: a receptive attitude toward even the most "primitive" peoples was manifested because such peoples were thought to represent some of the facets essential to the luster of the culture of mankind as a whole. The term "political-rational" has been applied to such cultural absolutism, but as Meyer has aptly said:

> Herder's concept of *Cultur* . . . contains seeds of both the political-rational and the irrational-cultural strands:[23]

> For . . . representatives of eighteenth-century enlightenment in Germany, the enlightenment itself . . . represented *Kultur,* and to emulate the achievements of *Kultur* was the task they set for Germany. *Kultur* thus had a universal, patently international flavor. Nonetheless individual nations or states could be regarded as the principal carriers of *Kultur,* and those nations were acclaimed as pathfinders for backward Germany. In this spirit, German radicals during the last decade of the eighteenth century supported revolutionary France and hailed Napoleon as the spreader of *Kultur* over all of Europe.

> The . . . [irrational-cultural] strand tended to regard *Kultur* as a complex of qualities, achievements, and behavior patterns which were local or national in origin and significance, . . . non-transferable, and non-repetitive. . . . The stress on such unique culture patterns . . . can be regarded as an attempt to compensate for a deep-seated feeling of inferi-

[23] Kroeber et al., *op. cit.,* p. 208, fn. 6.

ority on the part of German intellectuals once they had come in contact with the advanced nations.[24]

From this it is clear that although the German "political-rational" view of culture was superficially like French cultural absolutism, it lacked its tolerant, serenely confident quality. Moreover, what was earlier characterized as the French "delight in the quaint, the exotic, and the titillatingly unfamiliar" was in many ways different from that German "irrational-cultural" relativism to which Meyer refers. The consequences of such differences are hard to trace in German and French anthropology and sociology if they are regarded as self-contained academic specialties; clues must be sought in the more inclusive usage and eventual political fate of terms such as *Kultur und Zivilisation, Volk und Volksgemeinschaft, social, civilisation,* and so on.[25] Except for an occasional reference or allusion in later pages, however, it seems wise to devote no more space to what is after all an intricate problem of general intellectual history; it can safely be said here that French and German developments were distinguishably different, and that is enough.

Mention has already been made of the emphasis on culture in the plural manifested after the results of Alexander von Humboldt's journeys had become widely known. It may seem odd that so much importance should be attributed to a naturalist-traveler whose activities came as late as the nineteenth century, but the minor character of Germany's role in the era of European exploration and expansion must not be forgotten. For hundreds of years Spain, Portugal, Britain, France, and Holland had been ranging far and wide while Germany was huddling in parochial isolation. That isolation gave way to accessibility just at the time when *Kultur* in both the "political-rational" sense of the Enlightenment and the "irrational-cultural" sense of Romanticism became familiar to an enthusiastically literate public; the result was not only Humboldt's *Kosmos* in many editions but also a great increase in the publication of histories of culture. *Kulturgeschichte* of all kinds began to be favored by an ever more numerous swarm of book buyers. Cultures and peoples were dealt with from many standpoints: comparative folklore, dance, music, literature, art, technology, law, and the like were all the rage.

[24] *Ibid.*, pp. 207–208. Cf. Niedermann, *op. cit.*, pp. 216–218 *et passim*, with reference to Kant, etc.

[25] Some attention has been given to this search in Kroeber et al., *op. cit.*, pp. 28–29 *et passim*. Niedermann also refers to it in *op. cit.*, pp. 224–225.

Ethnographers and ethnologists contributed their share, but there can be no doubt that for a long time "philosophers, historians, and literary men were more active and influential. . . ."[26]

This was particularly true of the historians. As Small has pointed out,[27] the early nineteenth century in Germany was a period during which an overwhelming interest in historical documentation led to the collection of vast amounts of information about hitherto obscure topics and peoples. Great archives were established, historical museums founded, and a monumental series of publications begun. Greek, Roman, and Germanic antiquity received the most attention, and in the nature of the evidence revealed many conclusions were eventually reached that in other countries at other times came chiefly from the ethnologists. (Witness, for example, the vogue of Ruth Benedict's *Patterns of Culture* in the nineteen-thirties; Burckhardt and Nietzsche had drawn the Apollonian-Dionysian contrast a good three-quarters of a century earlier.) The impact of the massive documentary accumulations and their systematic utilization finally became apparent in the cultural relativism and functionalism of the German-Swiss classicist Bachofen, in what may justly be called the "personality and culture" school centering around Karl Lamprecht in the late nineteenth century, in the *tour de force* of Eduard Meyer's *Geschichte des Altertums,* with its ethnological introduction, in Kurt Breysig's wide-ranging albeit misguided social evolutionism, and in Max Weber's sophisticated handling of topics such as magic, *charisma,* and so forth. Granted, historical scholarship later played a fateful part in erudite aberrations such as Spengler's,[28] as did also the "irrational-cultural" conception of *Kulturen* as hermetically sealed wholes—but knowledge may be misused at any time or place.

The comparative interest also manifested itself among the philologists. Even though nearly a century-and-a-half has elapsed since the beginning of the era of the Grimms, Schlegel, Bopp, Lachmann, and Wilhelm von Humboldt, the enthusiasm of these researchers into languages previously unknown still passes readily from their musty pages to the reader. The final proof of the connections of the Indo-European languages, the realization of the importance of Sanskrit,

[26] Kroeber et al., *op. cit.,* p. 26.

[27] Albion W. Small, *Origins of Sociology,* Chicago: University of Chicago Press, 1924, pp. 37–109.

[28] This has been dealt with at some length in "Prospects of Social Change as Viewed by Historian and Sociologist," Ch. 3 of *Through Values* cited on p. 110.

the discovery of linguistic enclaves in Europe such as Finno-Ugric, and the revelation of the regularity of "sound shifts" made the early nineteenth century in Germany, and elsewhere in Europe too, a period that eventually became of the utmost importance for ethnography and ethnology, linked as they are with the study of language as one of the central phenomena of culture. From these fields of anthropology, attention to language eventually spread to social psychology and sociology, particularly among the symbolic interactionists and those interested in sociology of knowledge. Ernst Cassirer's work on "the philosophy of symbolic forms" is among the more recent examples; Wilhelm Wundt's studies of language in relation to ethnic mentality (*Völkerpsychologie*) represent earlier effects.

Along with influences stemming from naturalist travelogues, history, and philology went the very early impact of Biblical higher criticism. At about the same time—namely, the late eighteenth century—as Wolff was dissecting Homer, searching textual exegesis began to be applied to Holy Writ. There were sustained efforts to understand Hebraic "primitives" such as those in II Samuel, for example, and mere philological analysis proved insufficient. *Kultur* and *Volk* proved to be useful categories, although as one might expect, biological explanations depending on *Rasse* came into prominence as racialism gained learned audiences toward the middle and end of the nineteenth century.

This interpenetration of Biblical scholarship and what often amounted to anthropology, to the benefit of both, is hard to comprehend unless the striking difference between university organization in the German-speaking countries and that sometimes prevailing elsewhere is held in view. Most of the German-speaking universities had both Catholic and Protestant theological "faculties" (comparable to schools, divisions, or departments in the United States) as *integral* parts; the consequent connections with philosophy, the social sciences, and even the natural sciences were often quite close. Hence there was relatively little isolation of Biblical scholarship, herein affording a marked contrast to the set-apart theological seminary situation generally prevailing here. Not only did the anthropologists profit, as the Biblical sophistication of many of them showed, but so likewise did the sociologists, sometimes through the mediating function of the anthropologists, and sometimes directly. Max Weber's work in sociology of religion, for example, would have been quite impossible had knowledge of theological studies not been included as a matter of course in social-scientific training.

It was perhaps because of this close relation of theological studies with general university training that the battle about biological evolution, so bitterly fought in some countries, in the German-speaking amounted to only a few skirmishes. The most sweeping conclusions of physical anthropology encountered only slight and scattered opposition from the professors of theology or from the clergy as long as these conclusions were communicated chiefly to the university-educated and in the restrained, scholarly-scientific manner traditionally approved.

Antagonism was aroused only when Social Democrats and similar "fellows without a Fatherland" made political capital out of evolutionary doctrine and established Monist leagues having the Huxley-like Haeckel as patron saint. The historical materialism of Marx and Engels likewise formed the basis for a worker's *Weltanschauung* incorporating, before Kropotkin, a kind of social Darwinism in reverse; the class struggle was given biological interpretation, justification, and "expropriate-the-expropriators" solution.

German anthropologists and sociologists were for some time influenced by those more "authentic" versions of social Darwinism extolling the upper classes and the dominant peoples, and their effects on American sociologists such as W. G. Sumner and A. G. Keller are perhaps traceable—but the English-speaking and French-speaking countries also had their share of home-grown social Darwinists. Moreover, the influence on Sumner of German ethnologists such as Julius Lippert was definitely not of social-Darwinist type; what Lippert transmitted was intense conviction of the importance of culture.

In discussing naturalist-travelers such as Alexander von Humboldt, reference was made to the fact that the German-speaking countries played an altogether inconsequential and belated part in the expansion of Europe overseas. The result was that until well toward the last quarter of the nineteenth century German anthropology was strongly bookish and museum-oriented, with the books and the museums often stemming originally, where the sources were concerned, from the scholars, scientists, and field-workers of the non-Germanic countries. True, immense stores of firsthand materials about the more exotic peoples of the Dual Monarchy, Poland, Russia, the Balkans, Asia Minor, and similar nearby regions had been piled up by German historians, folklorists, and geographers, but these peoples were not then regarded as relevant for conventional ethnography and ethnology. The sociologists, however, were more catholic, and they drew heavily on such materials; descriptions and analyses of little-known European and European-fringe clans, communes, feuds, and the like that appeared in the late nineteenth and early twentieth centuries show striking resemblance to recent studies such as Dinko Tomasic's *Personality and Culture in Southeastern Europe.*[29] But the main point still remains: until Bastian, Ankermann, Ratzel, Preuss, von den Steinen, and other field workers from the German-speaking countries journeyed overseas, beginning after the middle of the nineteenth century,[30] firsthand anthropological studies were relatively rare.

Bastian and similar pioneers soon had followers, for with expansion under Bismarck and thereafter, German Southwest and East Africa, Samoa, New Guinea, and other regions afforded opportunity for extensive field work and discriminating analysis directly based thereon. Museum development was remarkably rapid. Frobenius, Graebner, Schmidt and Koppers, Thurnwald, and many other social and cultural anthropologists became internationally prominent. Their influence on German sociologists, however, was probably not as significant, even with regard to the problems of *Kultur* and *Volk*, as was that of the historians, philosophers, and philologist-linguists. Leopold von Wiese, for example, seems initially to have acquired much of his ethnographic knowledge from Herbert Spencer. Later, he leaned on Malinowski—but lightly.

If Wilhelm Wundt is classified as a social psychologist, and if social psychology is viewed as part of sociology, his ten-volume work on ethnic mentality (for this seems to be the best translation of *Völkerpsychologie*), saturated as it is with ethnographic data, may be thought to bring him among those influenced by the anthropologists—but his social psychology is at least ostensibly of psychological rather than sociological derivation. Tracing the channels presumably

---

[29] Becker and Barnes, *op. cit.*, Vol. 2, pp. 880–884, 1062–1065, 1078–1087 *et passim*. See also my *German Youth: Bond or Free*, London: Kegan Paul, 1946, Chs. 1 and 2.

[30] See Barnes, Becker, and Becker, *Contemporary Social Theory*, New York: Appleton-Century, 1940; use index for Bastian, Ratzel, and others. Naturally, R. E. Lowie's *History of Ethnological Theory*, New York: Farrar and Rinehart, 1937, provides authoritative comment.

flowing from anthropology to sociology hence becomes, in Wundt's case, a rather arbitrary procedure.

A more suitable example is provided by Alfred Vierkandt, an ethnologist who developed sociological inclinations. In close touch with the ethnographic evidence so rapidly accumulating in the late nineteenth and early twentieth centuries, he worked out a significant theory of cultural constancy and cultural change in his *Die Stetigkeit im Kulturwandel* (1908). In it he anticipated many of the valid conclusions found in Ogburn's *Social Change* (1923) and Wissler's *Man and Culture* (1923).

Vierkandt's central idea in this early work was that nothing in the realm of culture develops spontaneously, that everything is the product of gradual accumulation. In addition to setting forth the ideas of cultural constancy, continuity, inertia, and so on, he dealt at length with empirical evidence from technology such as that afforded by the development of the bicycle (here Ogburn's interest in invention was foreshadowed), as well as with evidence coming from economic institutions, language, art, and religion. He showed essential agreement with later cultural determinists, rightly or wrongly, by minimizing the role of the "great man" through calling attention to multiple inventions (again Ogburn) and similar phenomena. Further, Vierkandt showed awareness of "cultural lag" as characteristic of periods of rapid transition, particularly in modern times, although he did not erect on it his entire theory of cultural change nor infuse it with value-judgments.

Finally, Vierkandt attacked the mechanical theories of diffusion advanced by some of Ratzel's uncritical followers, pointing out that for any culture trait to diffuse there must be a certain readiness for its acceptance, and that, as Dixon later indicated, mere spatial proximity is a condition insufficient in itself to explain the transmission of culture.

Had Vierkandt continued to develop these significant ideas, instead of going all out for MacDougall's instinctivism, Husserl's phenomenology, and the holism of certain exponents of configurational psychology, he might have contributed greatly to German sociology and cultural anthropology. As it was, relatively little came of his efforts even among the sociologists with whom he later identified himself.

Another ethnologist with sociological leanings, Richard Thurnwald, now in his middle eighties but still with great energy as lecturer, writer, and researcher, and with the courage to have recently left the Communist-dominated University of Berlin and cast his lot with the Free University of Berlin in the American sector, is of special interest with regard to the relations of German anthropology and sociology.

Thurnwald began his career as a translator of the sociological writings of Lester F. Ward and Ludwig Gumplowicz, but soon shifted to ethnography, doing field work in Africa and New Guinea. Caught by World War I in the latter region, he was for some time interned in Australia, where he had opportunity to begin preparations for his monumental five-volume treatise on "human society in its ethno-sociological foundations" (*Die menschliche Gesellschaft in ihren ethno-soziologischen Grundlagen*); this reached completion in the early nineteen-thirties. Only one volume, that on primitive economics, has been translated, but that has already had important effects on American anthropologists such as Herskovits. His English articles on culture change, published in the *American Sociological Review*, seem to have attracted little attention.

Among Thurnwald's important earlier publications (1912) was a lengthy contribution to Gustav Kafka's handbook of comparative psychology, on the mentality of primitive peoples. In 1925 he followed up this "personality and culture" interest by reviving the old Lazarus and Steinthal journal of ethnic mentality (*Zeitschrift für Völkerpsychologie*), but adding to the title *und Soziologie*. After five or six years under this masthead, Thurnwald took the daring step of dropping the old title altogether and adopting a new one, *Sociologus*. The journal then published anthropological, social-psychological, and sociological articles and book reviews in several languages (chiefly English and German), and it circulated widely until 1934, when the Nazis forced an end to its publication. In 1953 *Sociologus* was revived, with Thurnwald still as editor—long may it flourish!

Still another German social scientist much influenced by anthropology is the sociologist Paul Honigsheim, formerly of the University of Cologne and Michigan State College and now in retirement in the United States. His many publications deal, among other things, with pastoral nomadism and totemism, with cultural zone (*Kulturkreis*) theories and their shortcomings, and with the ethnology of Bastian. Not yet known as well as he deserves to be, in part because of his bafflingly wide range of interests, sheer number of scattered articles, and break in career through Nazi persecution, he will eventually be recognized as one of the most important transmitters of anthropological knowledge to German sociologists. Much more should be said about him and other outstanding individuals, but space forbids; attention must now be shifted to the present situation in the German-speaking countries as regards anthropology and its relations with other social sciences and the public at large.

For many reasonably well-educated persons, anthropology is still identified chiefly with physical anthropology. It was earlier noted that this is also the case in the French-speaking countries, hence we cannot draw the inference that the German state of affairs is merely the result of Nazi racism or its antecedents. Instead of searching for the clue to the equating of anthropology with physical anthropology in racism or the like, it is probably best to draw conclusions from the language habits of those at a high level of literacy. For a long time anthropology was a general term for anything having to do with systematized knowledge of man; the "-ology" meant science only in the sense of *Wissenchaft* or *scientia*, not of natural science alone. Consequently, speculative teachings about man's "true" nature were, and today still are, permissibly entitled philosophical anthropology. Theological expounders of doctrines of original sin or of undefiled virtue refer to their particular brands of philosophical anthropology as revealing the essential nature of man. The following passage from a recent work by an American theologian much influenced by German and German-Swiss thought may justifiably be taken as representative of the usage noted:

> The complete contrast between the repudiation of Catholic optimism by the Protestant Reformation and the repudiation of both Catholic and Reformation pessimism about human nature in modern Protestantism is but one of many indications of the unresolved problems of Christian anthropology.[31]

Can it be wondered at if those skeptical of speculation think of anthropology without the qualifying adjective as physical anthropology, as the science of man's nature as dealt with by "hard-headed," non-theological natural scientists? Skin and bones and muscle seem to be such gratifyingly substantial materials for the study of man, and is not the old aphorism, *Der Mensch ist was er isst*, still current? In addition to this, *Volkskunde* and *Völkerkunde*, or even *Ethno-*

[31] Reinhold Niebuhr, *The Nature and Destiny of Man*, New York: Scribner's, 1941, Vol. I, p. 299.

graphie and *Ethnologie,* are more readily intelligible than *Anthropologie* trimmed with unfamiliar adjectives such as *soziale* or *kulturelle.* Mühlmann, to be sure, has recently published an excellent history of anthropology in which, in the title, *Anthropologie* is used in the American sense, but he represents an isolated instance, and it is highly unlikely that his example will soon secure a large number of followers. In any case, the text of Mühlmann's book is pervaded by the customary terms.

A further difficulty in communication for the anthropologist in the German-speaking countries arises from the survival of an antithesis that the present writer has hitherto mentioned merely in passing, and then only with reference to earlier French usage, "natural peoples vs. cultural peoples," *Naturvölker vs. Kulturvölker.* This is perhaps not desperately serious; English-speaking anthropologists have to wrestle with the layman's notions of "wild men," "natives in the raw," and so on, and lurid titles such as Malinowski's *The Sexual Life of Savages* are less than helpful. But when one finds the natural vs. cultural contrast deeply embedded in contemporary German social-scientific publications, as in the case of Vierkandt and others who should know better, that is somewhat disconcerting, to say the least.

Still another reason for the fact that anthropology has on the whole made a smaller net contribution to public information and to the other social sciences with regard to the implications of *Kultur* than have history, philology, and even Biblical studies is perhaps because, ever since the days of Alexander von Humboldt, Ratzel, and others, anthropology has been strongly tied to geography and the other earth sciences—geology and the like. Franz Boas, who miraculously enough later became one of the most dogmatic American champions of cultural determinism, was initially trained in Germany as a physicist and "natural science" geographer, conducting his first important research on the color of seawater. Anthropogeography as geographical determinism applied to man was long in vogue, only to be succeeded by *Geopolitik* as an application of all the sciences of social man, plus geography in the more limited physiographic sense, to the end of Nazi world conquest. Those deluded by such dreams have for the most part been rudely awakened or have gone to their eternal rest. Nevertheless, anthropology in Germany and Austria has been prevented from realizing a number of the potentialities for general social-scientific and popular usefulness that American anthropology, to name no other variety, has recently developed in high degree.

This may seem a discouraging end of a long story, but it is not the end. For one thing, anthropology in the German-speaking countries shows many signs of vitality. For another, it is well to remember certain encouraging parts of the story already told. Interdisciplinary contributions are to be welcomed from whatever point of the compass they proceed, and in many ways Pufendorf's *cultura*, Toussaint's *culture*, and Herder's *Kultur*, as transmitted by scholars and scientists in the humanities and the other social sciences, have brought results that in other countries have recently come in large measure from anthropology.

IV

In his charming and yet profound little book, *The Heavenly City of the Eighteenth Century Philosophers*, Carl Becker referred to "climates of opinion"; perhaps the metaphor may permissibly be extended into "climates of interest." At any rate, it is clear that when talk about natural man was widely current in the French-speaking countries (the German-speaking, as previous presentation may have implied, were somewhat less prominent), a like interest was to be observed in the British Isles; the intellectual weather seems to have been much the same.

For a number of reasons, not least among them the traditionally close connections of Scotland with France as a consequence of the Old Alliance, and with the Continent generally through the Wandering Scot as trader, mercenary soldier, student, and what not, the Scottish climate of interest offered a parallel even closer than the English to that of the countries across the Channel.

There were, to be sure, some queer gyrations of the Scottish barometer. Nevertheless, Lord Monboddo's persistent inquiries among Edinburgh midwives for babies born with tails was more than a mere eccentricity; the good old lord was nobody's fool, for he knew that human embryos at a certain point of development do have a discernible, but normally temporary, "prolongation of the lower spine." Even that illustrious group of social scientists so ably dealt with by Gladys Bryson in her *Man and Society: the Scottish Inquiry of the Eighteenth Century*—Adam Smith, David Hume, Dugald Stewart, Adam Ferguson, *et alii*—were anything but indifferent to the problem of natural man's biological characteristics. (It may be surmised, however, that their opinion was that Monboddo would have been better advised had he questioned the midwives of London, Cardiff, or Dublin.) But seriously as they took the question of "biologically natural man," that of "socially natural man" was taken still more seriously; it is a commonplace of intellectual history that they all rang changes on the latter theme.

Perhaps Smith and his fellows were inclined to deal with "socially natural man" not only because of their familiarity with writers such as Montesquieu, but also because of their keen awareness of Highland "primitives" who repeatedly made known their neighboring presence, although not their neighborliness. Lowlanders as well as Southrons were fair game for cattle-lifting clansmen from the "desolate and barbarous wastes." Moreover, there doubtless remained vivid memories of Montrose and Claverhouse and the Old Pretender and Bonnie Prince Charlie, who for all their Continent-tinged "civility" were viewed as Highland chieftains by their followers—"wild waved the eagle plume, blended with heather." They could command the fierce and unquestioning loyalty of barefooted Gaels who chanted the songs of Ossian while brandishing broadsword and target.

Still, the time when sentimentalizing about Ossian's mournful Celtic twilight was all the fashion had not yet fully arrived;[32] there was no exaltation of natural man by the Edinburgh *philosophes*. It may well be that eighteenth-century Scottish Calvinism, from the effects of which not even a Hume could wholly escape, had been less alloyed with softer Christian anthropology than had the contemporary Swiss version of the grim predestinarian doctrine. In any case, the former stood in even sharper contrast with Rousseau's notions of natural virtue than did the latter; some of the Scottish thinkers seemed to relish a certain harsh realism in description and analysis of the Highlanders—terms such as filthy, beggarly, cutthroat, overweening, and superstitious were freely used.

But after all, the Scottish share in the climate of interest should not be over-emphasized; everywhere in the British Isles, and particularly in England, there has long been much attention to folk-

[32] Macpherson began to publish fragments of his "translations"—which are today regarded as very free adaptations, but by no means as forgeries—in 1760, and the Ossianic collection came out in 1765. The full impact on the Romantics did not occur, however, until well toward the end of the century.

song and folklore, local history, fringe and enclave peoples, roving
tinkers and gypsies, and so on. Bishop Percy, with his *Reliques of
Ancient English Poetry* (1765), provides but one of the many possible
examples of antiquarianism, long antedating the Romantic vogue of
the past that took such an upsurge when the writings of Scott reached
the broad public. Societies of amateurs (in the etymological sense!)
sprang up and went happily to work gathering anything that was of
interest, and they eventually acquired almost professional skill and
organization.

A great deal of evidence was collected that was later to be profitably utilized by anthropologists and similar social scientists; indeed,
so much was amassed that some of it has not yet been worked over
thoroughly. A considerable amount was documentary, or at least was
rapidly recorded, as in the case of folklore. Naturally, another large
portion was archeological, but of a kind that reinforced and was reinforced by the written accounts, i.e., it was "historical" and more or
less directly intelligible. Along with this, however, archeological materials of prehistoric import, sadly in need of interpretation that only
later times could provide, were also assiduously accumulated.

The middle of the nineteenth century was drawing near before
those later times began, and the archeologists whose heyday then
opened were not so fortunate, where favorable prehistoric sites were
concerned, as were their fellow-workers in France. Nevertheless,
there were a few good spots to dig, and when to the data coming
from these were added the accumulations of the eighteenth-century
antiquarians and their successors, particularly in England, some of
the early phases of man's development began to be at least vaguely
outlined and understood. After the appearance of Darwin's *Origin
of Species* and *Descent of Man,* and especially of Huxley's *Man's
Place in Nature,* all such varieties of evidence, and the legitimate inferences from them, were scientifically interpreted and publicized
with great success.[33]

And not only scientifically: there was in late nineteenth-century
Britain perhaps more aggressive evolutionism, with a fervor ap-

---

[33] The best single reference, where completeness of detail is concerned, for
Britain from about this time onward, is T. K. Penniman, *A Hundred Years of
Anthropology,* 2nd ed., London: Duckworth, 1952.

For calling my attention to the second edition of Penniman, and for critical
comment on this entire section (IV), I am indebted to my colleague, C. W. M.
Hart, who, however, shouldn't be blamed for what I have done and left un-
done.

proaching the fanatical, than anywhere else in the world. The deep gulf between university education, on the one hand, and the hit-or-miss training of many Nonconformist class leaders, lay preachers, itinerant ministers, and local clergymen, on the other, made mutual understanding and compromise difficult if not impossible. As a consequence, among the poorly or partially educated, doubt resulted either in complete rejection of all religious tradition or a desperate clinging to the ultimately untenable or even absurd. Intolerant sects of Rationalists opposed intolerant sects of Bible-fetishists. Anthropology gained relatively little among the university-educated that would not have been gained anyway, and among the populace at large the ardent promulgation of generalizations held only tentatively by anthropologists, but dogmatically by Rationalists and their like, delayed by as much as a quarter-century the achievement of some measure of open-mindedness.

Slightly preceding in actual publication the evolutionism of Darwin-Huxley type was the more speculative variety represented by the sociologist Herbert Spencer, but this did not actually become widely known until after Darwin and Huxley had made their heaviest impact. Indeed, there is much warrant for saying that Spencer got his earliest general reader audience in the United States. In the eighteen-seventies John Fiske, on the lecture circuit, acted as his missionary, and in the same decade William Graham Sumner gave him some color of academic respectability (which he has never yet achieved in Britain) by making use of his writings for textbook purposes at Yale. But when once Spencerian evolutionism did reach the rank-and-file reader in Britain, its popularity was tremendous everywhere outside the universities, and even there it made inroads among the natural scientists.[34]

Much of the interest in anthropological publications at this time therefore stemmed from the interest stirred up by Spencer, more especially as he made extensive use of ethnographic materials in his writings. (Historical evidence, relatively little utilized in *Principles of Sociology*, was piled up in heaps in *Descriptive Sociology*, but he had no opportunity to profit from this tremendous compendium, and it remains almost unknown.) Professional anthropologists

---

[34] For careful and, for its time, surprisingly critical discussion of Spencer, see Leopold von Wiese, *Zur Grundlegung der Gesellschaftslehre*, Jena, 1906. See also Howard Becker, *Systematic Sociology on the Basis of the* Beziehungslehre *and* Gebildelehre *of Leopold von Wiese*, New York: Wiley, 1932, pp. 687–689, 710–712.

soon came to be disdainful of the travelers' yarns and other questionable sources on which Spencer, in his day and generation, had
been compelled so heavily to rely, and eventually their attitude was
conveyed to the sociologists. After World War I, to set the very latest
date, data were seldom if ever openly drawn from Spencer's treatises
or from the works to which they referred.

Some of his conceptual apparatus, however, did not suffer under
such heavy discount. Structure and function had long been household words among anatomists and physiologists, and Spencer's use of
them in his "principles" books on biology, psychology, and sociology
must have seemed eminently fitting to his readers, most of whom
were prepared to be impressed by terms having medical and natural-
scientific connotations. Further, structure and function fitted in
beautifully with the organismic analogy to the elaboration of which
Spencer devoted so much misguided effort.

Whether those British anthropologists who later explicitly called
themselves functionalists were directly influenced by Spencer is
doubtful; it is probable that Durkheim and such of his disciples as
Mauss and Hubert were of greater proximate importance. Yet, it
must not be forgotten that Spencer was favorably known to many
French ethnologists and sociologists adhering to the Comtean tradition, for the Englishman's organismic turn of mind elevated him, in
their view, to the level of the great author of the *Philosophie Positive.*
Durkheim certainly did not disavow his debt to Comte, was well
aware of Spencer's findings, and himself discreetly made use of organismic analogy (granting, of course, that he never perpetrated the
stupidities of a René Worms).[35] "From Spencer to Durkheim to
British and British-influenced functional anthropology to structural-
functional sociology in the United States" consequently may not be
a drastic distortion of the actual "who to whom" sequence. To be
sure, this is but one of the several channels leading into the main
stream of present structural-functional theory, but it is by no means
inconsiderable.

Reference has already been made to the assault by aggressive and
dogmatic evolutionists on the already crumbling but still formidable
strongholds of rigidly orthodox religion, and the consequent delay
in calm communication and understanding acceptance of attested
anthropological findings. Nevertheless, account must be taken of
the fact that there was a religiously influenced aspect of British anthropology in the nineteenth century. Missionary enterprise, along

---

[35] Becker and Barnes, *op. cit.*, Vol. 1, pp. 687–688, Vol. 2, pp. 829–838.

with the predominantly Protestant traits of British Christianity, produced even in university circles a strong interest in the *beliefs* of primitive peoples.

This is apparent, for instance, in Tylor's *Primitive Culture;* for all his scientific detachment and hard-headed respect for facts, his Quakerism led him to give a disproportionate amount of attention to animism as belief; ritual received relatively little notice.[36]

It is notorious that Spencer, with a clergyman for an uncle and familiarized by his upbringing with the doctrines of Methodism and Quakerism, was preoccupied with questions of religious belief ranging from the Unknowable of *First Principles* to the Euhemeristic ghost-theory of *Principles of Sociology.*

This religious interest was also manifest in the work of the Scottish classicist-anthropologist, James Frazer. His studies of the Greek and Latin literatures, together with his profound Biblical scholarship, led him to search not only in the sources indicated but also among "primitives" the world over for evidence bearing on the roles of magic and religion, magic and science, beliefs in immortality, the numerous accounts of dying and reborn gods, and so on. Most of his results were incorporated in his famous *Golden Bough,* but he published many other important and influential treatises.

Frazer's *Totemism and Exogamy,* for example, profoundly influenced Freud; on its unfortunately unstable basis the founder of psychoanalysis established his *Totem and Taboo,* a book thoroughly discredited among informed anthropologists the world over but still the cornerstone, visible or concealed, of genuinely orthodox Freudian instruction and practice. For a considerable period, now almost entirely elapsed among social psychologists and sociologists of standard academic training, Frazer was transmitted through Freud to a wide audience still numbering among its members not only some "intellectuals," many social workers, and a few younger anthropologists devoted to certain kinds of "personality and culture" doctrine, but also hordes of writers, readers, performers, hearers, and viewers of mass communications ranging from drugstore paperbacks to 3-D motion pictures. Who to whom, when and where and how, become quite complicated, to put it mildly.

But we are getting ahead of the story. Returning to Frazer and

[36] Tylor dealt with much, much more than belief, of course; if this were a history of ethnological theory (see Lowie, *op. cit.*), justice could be done to him. As it is, a seemingly cavalier brevity is imposed by present space limits and the nature of this symposium.

continuing to call attention to his interest in problems of belief, his little-known *Psyche's Task* should be noted. This book, a carefully reasoned defense of the role of superstition in social stability, fell stillborn from the press. In the effort to get its circulation started, it was rechristened *The Devil's Advocate*, but it never really came to life. This is the more regrettable because of its importance for the understanding of Frazer and his times, as well as for the interesting parallels it shows with the thinking of the French conservatives, De-Bonald and DeMaistre, and with aspects of Durkheim's work.

Perhaps in part because of Frazer's conservatism and his unimpeachable classical credentials, anthropology in practically all its branches soon became fully respectable and capable of integration in the offerings of even the traditional British universities. Frazer himself did little, however, to foster its application to the many problems within the Empire and Commonwealth toward the solution of which it might have contributed even more notably than was eventually the case. In fact, Frazer remained a book-anthropologist; he never did any field work. On one occasion, so runs the story, when questioned at a formal dinner by an earnest lady as to his possible eye-witness acquaintance with some of the odd and disturbing peoples and customs described in his writings, he exclaimed, "Madam, Heaven forbid! I want nothing to do with such creatures or their ways."

Fortunate it was for the social sciences in Britain that anthropology did become respectable, by whatever means, for sociology, alas, did not; even the great liberal, L. T. Hobhouse, could not transfer his *Manchester Guardian* prestige to his sociological writings. To this day, in fact, sociology labors under handicap. Its lack of respectability is in part the result of its very name. Comte boasted that in coining the term *sociologie* for his new science, by combining Latin and Greek roots, he thereby bestowed upon his beloved France, only legitimate claimant to the heritages of Greece and Rome, incontestable evidence of intellectual supremacy. Sociology stood at the pinnacle of the sciences; France headed the concert of nations; consequently . . . Well, even under the *Entente cordiale* this sort of thing could elicit only raised eyebrows in Britain. "Really, such—ah—bastard Latin-Greek etymology had better be passed over in silence."

Moreover, sociology sent forth faint overtones seemingly identifying it in some respects with socialism. Perhaps the identification was in certain respects warranted; Shaw has somewhere said that the early Fabian Socialists knew their Comte far better than they knew their Marx. However this may be, the work of the Webbs certainly did much to identify sociology with "social science," that intensely practical British equivalent of a combination of American "social work" and "public welfare administration." Even today memories of such supposition of affiliation demonstrably linger.

Also relevant here is what has previously been said about Spencer and evolutionism; he and his teachings had been adopted by the Rationalists and similar undignified persons. To cap it all, his book, *Education,* had alienated many humanists because of its championing of the natural sciences, and the humanists had much to do with the administration of the traditional universities. "Sociology? Oh, yes, yes. Now what were we saying?"

As a result of the various trends mentioned above, and many more, it is readily understandable why it was that social anthropology, having a number of characteristics that in other countries would be called sociological, and occasionally were in Britain, was what eventually appeared in strength in the British academic theater. The great age of social anthropology may be said to have dawned at the very beginning of the twentieth century with the work of Rivers and Haddon at Cambridge (although Westermarck, after 1903 at the University of London, finally occupying one of the Martin White chairs of sociology there, was from some standpoints an ethnologist, and had made a considerable stir with his *History of Human Marriage* in the early eighteen-nineties). In due time anthropologists such as Seligman, Malinowski, and others at the University of London achieved prominence, and concurrently Radcliffe-Brown,[37] virtually a Durkheimian disciple, began to make what eventually became a very strong impression. Oxford, Cambridge, and Edinburgh lent more and more of their great prestige to anthropological studies, and today the names of social anthropologists loom large on the British intellectual scene—Evans-Pritchard, Forde, Schapera, Nadel,

[37] Radcliffe-Brown is difficult to localize. First exposed to Cambridge in the Rivers-Haddon days, he studied for a time with Durkheim at the Sorbonne, then went to the Andaman Islands to do the field work that resulted in his famous book, then undertook another field work venture in Australia that was interrupted by World War I. From 1921 to 1926 he was professor of anthropology at the University of Cape Town, then at the University of Sydney until about 1931, then at the University of Chicago until 1937, and then, as the first professor of anthropology since the days of Tylor, at Oxford, where he retired in 1946. He was also visiting professor at Yenching, Cairo, and São Paulo. In spite of all this moving about, he doubtless has done more to found a distinctive school of British social anthropology than has anyone else.

Firth, Fortes, Hogbin, Curle, and many more. So many and so able are they that they should soon begin to exert revivifying influence on the British sociologists, who although by no means moribund— as witness the *Sociological Review* and especially the newly-established *British Journal of Sociology*—need aid, comfort, and recruits.[38]

In the summer of 1953, British social-anthropological enterprise was exemplified by the launching of researches on Shropshire, preeminently a rural county, that could well have been carried out by sociologists, especially those trained in the rural field, had there been any sufficient number of such sociologists with the requisite skills and interests. As it is, up until now local historians, churchmen, administrators for this or that governmental agency, land economists, geographers, and so on, have monopolized the study of the countryside (the English "place-work-folk" followers of Demolins represent lone exceptions). The social anthropologists should start a period, if not of "good-bye to all that," at least of effective utilization of existing resources.

It is likely, though, that the sociologists will not only continue to recover from the near-paralysis from which they suffered during most of the second quarter of this century, but will progressively increase their speed of convalescence and their utilization of the remedies offered by sociologists of other countries. (In presenting, on the next page or so, certain aspects of a three-cornered discussion by Fortes, Murdock, and Firth with regard to the essential features of current British social anthropology, this matter of relations with British sociology will come up again.)

The British sociologists have learned something, however, and can learn much more, from the results of social anthropology as applied to the problems of the Commonwealth and Empire. British expansion in the nineteenth century, and contraction in the twentieth, involving every continent and Oceania as well, brought with it a maze of problems and at least provisional solutions that can be roughly indicated by phrases such as "the white man's burden,"[39]

[38] See Becker and Barnes, *op. cit.*, Vol. 2, Ch. 21, and 1937–1950 Appendix.

[39] It is today fashionable, in some circles, to be a trifle cynical about this kind of thing, but compared with the record of *any* European or American people *vis-à-vis* the inhabitants of colonies, dependencies, mandates, and so on, the British make a distinctly creditable showing. But these comments are perhaps extraneous in this context.

indirect rule, utilization of local manpower resources, colonial administration, official and unofficial advisory functions, and the like. Anthropology, once it had acquired respectability, was rapidly included among the social sciences that could be advantageously applied.

This was particularly true of social anthropology, which in Britain is now the preferred term for at least certain aspects of what is frequently called cultural anthropology in the United States. The British leaning toward social anthropology, which deals chiefly with matters such as kinship systems, religious practices, and political regulation, rather than with the full sweep of culture, may have arisen primarily from the fact that after resort to resident advisers and indirect rule became more and more a necessity to an imperial power with only "a thin red line" and suave persuasion at its disposal, anthropological field workers paid most attention to problems of control. The colonial administrator needed first of all to know the intricacies of social organization; the less obvious or less politically central aspects of the culture might be taken into account *after* proper ruling procedures had been established. It does not take much reading between the lines of the following statement *re* social anthropology in Britain by Meyer Fortes, one of the leading British anthropologists, to see the relevance of the above comments about political problems:

. . . Most social anthropologists would now agree that we cannot, for analytical purposes, deal exhaustively with our ethnographic observations in a single frame of reference. We can regard these observations as facts of custom—as standardized ways of doing, knowing, thinking, and feeling —universally obligatory and valued in a given group of people at a given time. But we can also regard them as facts of social organization or social structure. We can then seek to relate them to one another by a scheme of conceptual operations different from that of the previous frame of reference. We see custom as symbolizing or expressing social relations—that is, the ties and cleavages by which persons and groups are bound to one another or divided from one another in the activities of social life. In this sense social structure is not an aspect of culture, but the entire culture of a given people handled in a special frame of theory. . . . And no doubt as our subject develops other special techniques and procedures will emerge for handling the data. No one denies the close connections between the different conceptual frames. . . . By distinguishing them we recognize that different modes of abstraction calling for somewhat different emphases are open to us.

The recent trend in British social anthropology springs primarily from field experience. . . . A prominent feature . . . is the attention given to the part played by descent rules and institutions in social organization, and the recognition that *they belong as much to the sphere of political organization as to that of kinship.*[40]

Murdock's remarks about British social anthropology were made quite independently of those just quoted, but they read almost like a commentary thereon:

The British social anthropologists, in the first place, do not concern themselves with the entire range of cultural phenomena, but concentrate exclusively on kinship and subjects directly related thereto, e.g., marriage, property, and government. To be sure, some of them . . . have dealt with economics, and others . . . with religion. Nevertheless, it is an incontrovertible fact that such major aspects of culture as technology, folklore, art, child training, and even language are almost completely neglected.

A second limitation is geographical. *For a generation hardly a single professional British ethnographer has worked with any society not located in a British colonial dependency.* . . .

A third limitation, related to the foregoing, is an almost complete disinterest in general ethnography—difficult to account for in a country that has produced a Tylor and a Frazer. Of the two or three thousand primitive societies in the world whose cultures have been recorded, the British social anthropologists as a group reveal a concern with and knowledge of not more than thirty. . . .

Sober reflection has led . . . to the startling conclusion that they are actually not anthropologists but professionals of another category. . . . The special province of anthropology in relation to its sister disciplines is the study of culture. *Alone among the anthropologists of the world the British make no use of the culture concept.* Assuming culture to be their province, most anthropologists feel free to explore its every ramification. The British school alone concentrates upon a few words of Tylor's classic definition and rules the rest out of bounds, including such aspects as technology and the fine arts.

. . . the only claim of the British school to the name of anthropology rests on the fact that they conduct much of their field research in nonliterate societies.

In their fundamental objectives and theoretical orientation they are affiliated rather with the sociologists. Like other sociologists, they are interested primarily in social groups and the structuring of interpersonal relationships rather than in culture, and in synchronic rather than in diachronic correlations. . . .

Though unmistakably to be classed as sociologists, the British social anthropologists should not be associated with contemporary sociology, which has absorbed so much from both psychology and anthropology that

[40] Meyer Fortes, "The Structure of Unilineal Descent Groups," *American Anthropologist,* 55, 1, Jan.–March, 1953, pp. 17, 21, 23, italics not in original.

it has become almost indistinguishable from the latter in its fundamental theoretical orientation. The comparison should rather be with the sociological schools of an earlier generation . . . e.g., Sumner, Pareto, and Thomas, whose theories were current in the 1920's.[41]

To Murdock's assertion that the recent "who to whom" in Britain has apparently been from an earlier and now somewhat superseded sociology to a sadly limited social anthropology, Firth has made a direct reply:

Murdock . . . points out that what British anthropologists are doing is essentially a specialized kind of sociology. He is startled, he says, to come to such a conclusion. It is hard to know whether this is a case of judicial ignorance or of magical fright. For he has been well warned. The position as British social anthropologists themselves see it has been explicitly as well as implicitly stated on a number of occasions.[42] Moreover, their relations with sociologists such as Hobhouse, Westermarck, and Mauss were close at a formative period of the science, as also later. *The more general theory of the anthropologists, then, is hardly distinguishable in its scope from that of the professed theoretical sociologist,* though its different ethnographic base gives it a different illustrative content and a different—sometimes sharper—focus.

If the classification of social anthropologists as sociologists is conceived merely as a convenient way of labeling them, it is of little moment. But it is significant if it emphasizes that their primary connections are not with the human biologists who study physical anthropology, nor with the students of primitive technology who are concerned with embryonic aspects of applied mechanics, nor with the archeologists, whose major role as I see it lies with the historians. . . .

The primary connections of the social anthropologists are with the other social scientists—in sociology in the narrow sense, in psychology, in economics, in political science, in jurisprudence, and even in such history as is problem-orientated. . . . *What is relevant here is the aim of strengthening linkage between disciplines, and not of simply making a case for an old-fashioned—and spurious—unified science of man.*

As to the charge of neglect of culture, . . . Fortes has argued that "culture" and "structure" denote complementary ways of analyzing the same social facts. . . .

My own view . . . follows much the same lines. . . . "Society" emphasizes the human component, the people, and the relations between them; "culture" emphasizes the component of accumulated resources, non-material and material, which the people through social learning have acquired and use, modify, and transmit. But the study of either must in-

---

[41] G. P. Murdock, "British Social Anthropology," *American Anthropologist,* 53, 4, Part I, Oct.–Dec., 1951, pp. 467, 469, 471, 472, italics not in original.
[42] See Firth's detailed footnote.

volve the study of social relations and values, through examination of human behavior.[43]

In effect, Firth says that Murdock is by no means wholly wrong as to the influence of earlier sociologists on British social anthropologists. (Here again "who to whom" is by no means simple.) What Firth does *not* say, however, is that contemporary British sociologists who are so labeled exert very little influence; those who do are French, German (Max Weber is getting a very respectful hearing), and American—even a hasty glance at the references scattered through the just-quoted articles by Fortes and Firth will show this. Murdock is wrong, then, about limitation to "sociological schools of an earlier generation," but right about the scarcity of reference to influential and up-to-date British sociologists.

But what of contemporary British sociologists who are not labeled as such? And what of the hearing accorded them before audiences larger and more inclusive than the specialized ones just discussed?

Prominent but unlabeled sociologists were probably more numerous in the generation immediately past than in the present;[44] today about the only one who has attracted much attention is Arnold J. Toynbee, ordinarily called a historian both by himself and others, but occasionally referred to as a sociologist by British historians—and in the still surviving derogatory sense. The reason for the name-calling undoubtedly lies in their conviction that it is impossible validly to generalize on the basis of historical evidence. Whether this is a correct conviction need not at this time detain us. However, one of the assumptions underlying the conviction—namely, that Toynbee is trying to generalize—is clearly correct. In this respect he can be placed in the same bracket as Kroeber, the American writer who has so strongly insisted on the study of culture as the *sine qua non* of anthropology, and who says at the beginning of his elaborate study of the rise and decline of large-scale culture patterns:

> That this book dealing with data from history should have been written by an anthropologist will perhaps seem fitting to those interested in the development of the two studies. *The aim of the work is obviously more or less sociological.* The principal current of anthropology, and its soundest

[43] Raymond Firth, "Contemporary British Social Anthropology," *American Anthropologist*, 53, 4, Part 1, Oct.–Dec. 1951, pp. 477–478, 483; italics not in original.

[44] See Becker and Barnes, *op. cit.*, Ch. 21.

findings until now, I believe to be culture-historical. Nevertheless, if we can also generalize validly, it will be most important.[45]

Little influence from British or any other anthropology—insofar as in one respect it can be identified with the professional study of nonliterate peoples—is to be found in Toynbee, and whatever traces are to be found are ethnographic and archeological rather than the result of direct acquaintance with social anthropology. Even the traces mentioned have to do only with certain of the "arrested civilizations"; the Polynesian, the Eskimo, and the Nomad (these are Toynbee's terms), and derive from only a relatively slight acquaintance with the source materials, most of them secondary. Toynbee has been given some attention by professional sociologists, as recent articles in the *British Journal of Sociology* testify, but all in all his audience is of broadly academic and/or high-level general reader character. Such an audience may be of considerable size, for the extensive sales of the massive *A Study of History,* to say nothing of the demand for the one-volume Somervell abridgment, would indicate as much—as is also shown, incidentally, by the interest in Sorokin's *Social and Cultural Dynamics,* of which a one-volume adaptation by Cowell has recently appeared.

But even when the domestic "unlabeled sociologists" and those foreign allies who follow recognizably similar lines of thought are included in the British sociological roll-call, it is still safe to say that the social anthropologists probably have a wider popular appeal. For example, Evans-Pritchard recently broadcast the substance of his somewhat technical book on social anthropology to an appreciative BBC audience.[46]

Beyond this, indeed, at the level of policy rather than popularity, it is probably only the economists and the historians (who often lecture on "politics" as well) who have greater academic influence on topflight policy-makers than do the social anthropologists. One might

[45] A. L. Kroeber, *Configurations of Culture Growth,* Berkeley and Los Angeles: University of California Press, 1944, p. vii, italics not in original. In a footnote on p. 834, Kroeber says this: "The works of Toynbee and Sorokin have pertinence here, in that they also are synthetic interpretations of broad masses of historical data. They did not come to my notice until the present study had been written." Toynbee's *A Study of History* appeared, where the first three volumes are concerned, in 1934; the second three came off the press in 1939. Sorokin's *Social and Cultural Dynamics,* first three volumes, bear the date of 1937. See also *Through Values,* etc., previously cited on p. 110, for discussion of Toynbee and Sorokin, pp. 149–154, 180–185.

[46] Although, to be sure, on the Third Program!

think, given the Welfare State features of British life persisting regardless of whether the Conservative or the Labor Party is in power,
that representatives of the amalgam of social work and public welfare administration known as "social science" (and to which reference has already been made) would be sought after as advisers. This
is to some extent the case, but not notably so, for much "social science" is an affair of on-the-job training or is linked with business
education (Commerce and Social Science is a not infrequent combination), and consequently enjoys little prestige.

Summing up, "who to whom" in Britain is an involved problem,
but is easier of solution for the present than for the past. Even if the
task assigned for this section of the chapter had been other than what
it is—namely, the tracing of the effects of British anthropology on
British sociology—it would still have been necessary to say that
today the anthropologists are to be numbered among the blessed
who give rather than receive.

What will happen tomorrow is any man's guess, but one can
readily imagine a sociological Dundee singing "Ye hae no seen the
last o' my bonnets and me." And such imagination implies no tacit
acknowledgment that sociology in Britain is a Lost Cause!

V

Anthropology in other than the loose sense noted above, in accordance with which it is applied to any intellectual endeavor,
speculative or otherwise, having to do with man, got under way
rather slowly in those parts of North America that eventually became
the United States.[47] The early missionaries, Roman Catholic and

[47] Given the fact that all the contributors to the present symposium received
at least the preponderant part of their training in the United States, that many
and perhaps most of our readers are somewhat familiar with American anthropology or sociology or both, that recently two popular books (Kluckhohn's and
Chase's) have dealt with the theme of this section, and that Lowie's history of
ethnological theory has a good deal to say about American anthropology up to
1937, the remarks made here in limited space may appear presumptuous as well
as superfluous. (The writer is keenly aware of this possibility because, for over
a decade, he has supplied for the *Encyclopaedia Britannica Book of the Year*
an annual article on developments in sociology, and in this article account has
always been taken of outstanding anthropological contributions, particularly in
the United States. These articles have recently been reprinted, with suitable
revision, as parts of the 1937–1950 Appendix on Sociological Trends in Becker
and Barnes, *op. cit.*, Vol. 2.) Nevertheless, it is hoped that even old ideas may
seem to have a tinge of freshness when presented outside their customary contexts.

Protestant, were of course very active in description, as were also some of the traders and those explorers who, like Mackenzie and Lewis and Clark, had a fair amount of peaceful contact with native informants, but their accounts were necessarily biased and fragmentary. (*The Jesuit Relations* provide only a partial exception, and in any case their effect was primarily on the French-speaking world.) Dispassionate interest in the Indian as such was rare. Even as late as the middle of the nineteenth century, vocational opportunities affording scope for such interest were so limited as to be virtually non-existent, although government service in connection with Indian affairs was eventually of some help. Except for Parkman's spirited descriptions, Morgan's justly famous books of social-evolutionary bearing, the little-known reports of the Bureau of Ethnology, and a few miscellaneous items, reasonably full and accurate anthropological publication was decidedly scarce; most of it was anecdotal, sketchy, and loaded with errors.

This is odd, in a way, for from the time of the earliest settlers onward there had been considerable interest in natural man. As might have been expected under the circumstances, the Negro sometimes provided the exemplar, but far more often and certainly far earlier, the Red Man was thought to be the best representative of man *ab origine*. Adam Ferguson, whose writings were by no means unknown in the colonies, once made the comment, *contra* Rousseau's lucubrations about the state of nature, that what was really needed was close observation of the Indians, not flights of fancy. Ferguson's searching and unemotional type of analysis was uncommon, however; the natural man of the New World was usually held to be virtuous or scoundrelly according to the convictions of those who depicted him.

Most of the time the Indian was believed to be a scoundrel, for the convictions were overwhelmingly to the effect that the settlers represented the righteous and the natives the unrighteous. W. C. Macleod's *The American Indian Frontier* graphically describes how the righteous went to work after they rose from their knees; "the only good Indian is a dead Indian" epitomized the prevailing belief and practice. Admiration for Hiawatha was an aberration of an effete Easterner who had absorbed too much of the doctrines of Herder and like Romantics. It was conceded that the Indian was superior in his sensory equipment, but this only demonstrated his animality.

Doubts arose, of course—Cooper fostered many. Leatherstocking was as close to nature as Chingachgook or Uncas, yet he sometimes seemed to lack the elevation of soul that distinguished the Mohicans. But the push to the West held doubts in check, and only after most of the forlorn aborigines had literally reached the end of the trail and were safely penned in reservations could their conquerors luxuriate in sentimentality. Belated Rousseaus turned up everywhere; caterers to current taste, from the barkers for the Wild West shows to the popular novelists, found that the Noble Red Man was a saleable commodity.

It would be gross distortion of the actual state of affairs directly to link these trends among the American populace at large with the more or less clearly scientific and scholarly interests that were beginning to find institutional support as American universities (as distinct from the colleges) developed beyond the traditional combination of law and medicine. Nevertheless, the fact that there was widespread popular backing for the anthropological exhibits at the Columbian Exposition and its successors, that great new museums featuring anthropology were established and old ones expanded and appropriately arranged, that publishers became willing to take risks on books that dealt with anthropological themes in scientific fashion—this was not pure chance. The climate of interest had become generally favorable to anthropology.

Sociology enjoyed a similarly favorable climate, and began to flourish even earlier—much earlier and more vigorously, in fact. This was partly because of the immediately practical traits, as practicality is ordinarily judged, that have been regarded as characteristic of sociology.

Comte followed the *Positive Philosophy* with its practical application, the *Positive Polity*, and in the very year that the last volume of the latter came off the press—namely, 1854—the first books in English bearing the term "sociology" on the title page likewise appeared. The purpose for which these books were written was intensely practical, for they were both determined defenses of the slave-holding as against the wage system.[48] Like Comte, the authors, Hughes and Fitzhugh, made much of the organismic analogy, although the associated concepts of structure and function received little attention. The collapse of American slave-holding doomed the

[48] L. L. and Jessie Bernard, *Origins of American Sociology: the Social Science Movement in the United States,* New York, 1943, pp. 7, 235, 249, 411.

books to an oblivion from which they have only recently been rescued.

More important than this early Comtean influence, in the establishment of the conception of sociological practicality, therefore, have been organizations such as the American Social Science Association, which began to hold philanthropic meetings in the eighteen-sixties. Here the term "social" was promptly equated with "sociological," with the result that for a long time one highly influential part of American sociology was considerably like today's British "social science."[49] What the British call the Three D's of social science, Drink, Drains, and Divorce, was early paralleled by the American Three S's, Sin, Sex, and Sewage. At present the establishment of separate schools of social work in many American universities has notably diminished this emphasis in sociology. Still, the ubiquitous and useful "social problems" courses, together with attention to juvenile delinquency, criminology, penology, and so on, help to account for the popular belief that a sociologist is like Joe Jefferson's character who "hangs around slaughterhouses, and drinks blood, and spits red, white, and blue." Most sociologists, however, are seriously concerned with building a firm empirical and theoretical base for their science, and abjure hard-guy exhibitionism.

The building of the scientific base can be said to have begun in the eighteen-seventies; Spencer's effect on Sumner has already been mentioned. Ethnography, albeit initially of no remarkably high quality, was consequently incorporated in the very foundations of American sociology, and as time went on a thoroughgoing cultural relativism drawing on Lippert[50] and many other reasonably accurate exponents of *Kulturgeschichte* and similar interests was erected. *Folkways* was but one of several such relativistic sociological treatises; its general influence, however, was far wider than that of any other—perhaps because of the social Darwinism it so convincingly recommended.

Only a couple of years after Sumner had begun to use Spencer's writings for textbook purposes, Lester F. Ward's *Dynamic Sociology* came off the press. Instead of social-Darwinistic and Spencerian op-

[49] *Ibid.*, Part VIII.

[50] Julius Lippert, *The Evolution of Culture*, trans. and ed. by G. P. Murdock, New York, 1931. For a cogent treatment of culture in general, showing clearly the reciprocal influence of sociology and anthropology, see Murdock's "The Science of Culture," *American Anthropologist*, 34, 2 April–June, 1932, pp. 200–215.

position to political intervention and control, Ward's work was essentially an eloquent appeal for State socialism, with more of Bismarck than of Comte among its intellectual sources. Its scientific apparatus was borrowed from the natural sciences in amount much greater than Sumner's, but a substantial quantity of ethnographic evidence, fairly accurate for its day, was offered throughout.

Ross, Small, and several other mildly left-wingish sociologists transmitted ideas about social control, class struggle, and governmental planning closely akin to or derived from Ward's. In Ross's case, however, Tarde and Durkheim were also of much importance; Small was profoundly affected by Marx as diluted by the German "professorial Socialists"; and neither of them made use of other than second-hand ethnography—when they used it at all.

In fact, the first American sociologist to utilize the work of the reliable ethnographers and ethnologists in a genuinely scientific manner was W. I. Thomas. His *Source Book for Social Origins,* published in 1909, was far in advance not only of what was being written or systematically collected by most other sociologists anywhere in the world, but was also considerably ahead of what many anthropologists were providing their readers, and this continued to be true, *mutatis mutandis,* of his later books. In Thomas's writings can be seen the skilful interrelating of prehistory, accounts of the non-literate peoples, the Greek and Latin classics, history over a surprisingly wide span, and knowledge of modern life in many countries in intimate detail. American anthropologists have paid little attention to Thomas, but much of what some of them have been excitedly heralding (with frequent changes of program) as the scientific advance of year-after-next, he had soberly expounded nearly a half-century ago. But why refer disparagingly to the anthropologists? The sociologists have been even more remiss, with less excuse.

Thomas's sophisticated use of anthropology was by no means a matter of chance. Shortly after the American climate of popular interest became favorable, there appeared on the academic scene, in 1889, the first professor of anthropology in any American university, Franz Boas. Born and trained in Germany, acquainted with Bastian and Virchow, he also did his first research, field-work, teaching, and museum classification under German auspices. Beginning his ethnographic data-gathering with the Eskimo, he then turned to the Indian tribes of the northwest Pacific coast; the latter venture brought him into contact with Tylor and, to that extent at least, with British anthropology. His first American university position was at Clark, but before long he located at Columbia, and was intensely active there for forty years, retiring only in 1936. To call the roll of his students would be to mention most of the American anthropologists well known to the present generation; his influence has been tremendous.

Boas was first of all a field-worker, with the field-worker's healthy distrust of "sweeping generalizations." To this, quite as much as to the influence of the anti-evolutionist Bastian, can probably be attributed his painstaking recording in the language of the people being studied, accumulation of historical evidence whenever possible, resistance to geographical or racial determinisms, predilection for viewing all conduct in full context, insistence on the importance of culture, *and* irritating unwillingness to systematize the theoretical leads he was perpetually throwing out. His attacks on the patron saints of social evolutionism—Morgan, McLennan, and their like—can best be understood from this standpoint. He went much too far in the direction of the merely idiographic and particularistic, but he did clear away many encumbrances. Moreover, even his particularism served a useful purpose, for it was so rigorous in method that it still serves as a reprimanding example to those who think that "description is easy."

Thomas said little about Boas's influence on him, but it is quite obvious to those who know the writings of both men. Directly acknowledging the salutary guidance of Boas and those sharing his high standards[51] was W. F. Ogburn, who perhaps did more than anyone else to familiarize his fellow-sociologists with the concept of culture.

---

[51] Here may be mentioned A. L. Kroeber, who has perhaps done more than any other anthropologist to keep the concept of culture in the forefront of attention, as witness his *Configurations of Culture Growth,* Berkeley and Los Angeles, 1944, and the collection of his important early articles, *The Nature of Culture,* Chicago: University of Chicago Press, 1952. Kroeber's primary importance, however, has been as an anthropologist's anthropologist; he has had less effect on sociology than many quite minor figures. The similarly eminent R. E. Lowie published, in 1917, a small book that made a profound impression, *Culture and Ethnology.* One might go on to list many others, as for example the sensitive, mercurial, and erudite A. A. Goldenweiser, who collaborated with Ogburn as co-editor of the significant symposium, *The Social Sciences and Their Interrelations,* New York, 1927, whose *Early Civilizations,* New York: Crofts, 1922, was outstanding for its clear presentation of a number of social and cultural systems, and whose chapter, "Contributions of Anthropology," in Barnes, Becker, and Becker, *op. cit.,* is a masterpiece of relevance and condensation.

Of course, as noted earlier, the substance of the concept had long been taken for granted by numerous historians, comparative philologists, and other American exponents of the humanities and the social sciences, particularly those trained abroad. In addition, "the cake of custom," "the social heritage," "folkways and mores," and many other terms and phrases of similar import were current among sociologists.[52] Nevertheless, the educational orientation of American sociologists was even then almost diametrically opposite to the readily available sources of systematic knowledge about culture— few of them really knew any history worth mentioning, for example. If they were to remedy their deficiencies, it would have to be through a discipline possessing prestige because of its connections with the natural sciences and offering presumably authentic information about biologically and socially natural man. Thomas had familiarized sociologists with the accurate descriptions provided by Boas and others; Ogburn not only continued this emphasis but also added the stress on culture.

It may well be that both Thomas and Ogburn found their tasks made easier by the fact that the period between the beginning of the twentieth century and the establishment of the quota system in the middle nineteen-twenties was one in which a major sociological preoccupation was with problems of immigration and assimilation— an aspect of what has recently come to be called acculturation. From 1900 to 1910, for example, the United States received more than eight million newcomers from overseas, and how these "one in every dozen" were to be suitably absorbed into American life was discussed on every hand. "The melting pot" idea gained currency rapidly, particularly among those sociologists who possessed what might be called home missionary proclivities—and many, at heart Protestant clergymen who had chosen to preach a secular gospel, did possess them. From hope for the efficacy of "the melting pot" to faith in the power of "environment" as over against "heredity" to the belief in the omnipotence of culture—once the concept had been clearly presented—was a sequence readily followed by many. The end-point reached may have been scientifically warranted, but the route was often far from direct.

During the time when Boas's influence was at its peak at Colum-

---

[52] See the bibliographies in Murdock's "The Science of Culture," cited on p. 145, and in Leslie A. White, *The Science of Culture*, New York: Farrar, Strauss, 1949, pp. 425–535.

bia, Ogburn took his graduate training there—but as a sociologist, not as an anthropologist. In sociology, Giddings was then making a strong case for the use of quantitative methods, and Ogburn was duly impressed. Combining his interest in culture with his statistical skills, he began to investigate multiple inventions, and hence developed a strong interest in technology as the all-sufficient factor in social change. This in turn led to a kind of material-culture determinism in many respects indistinguishable from vulgar Marxism, and the alliterative lilt of "cultural lag" soon served as supposed explanation and condemnation of anything thought undesirable by the sociologist concerned.

In spite of such drawbacks, however, Ogburn's function as transmitter of the stress on culture must be viewed as having been very much worthwhile—indeed, as indispensable. Sumner, with his emphasis on values and value-systems as the core content of culture, was sounder in many ways and had at least a thirty-year head start, but he didn't popularize the magic word.

As a result of functionalist and other influences, many of them emanating from Malinowski and those sharing his viewpoint, divergences from the Boas type of sternly scientific anthropology began to appear even before his prestige had reached a peak in the middle nineteen-twenties. Among the many evidences of this divergence was the "problems" approach. Paradoxically, Boas himself may be said to have helped initiate this through his *Mind of Primitive Man* (1911) and his studies of changes in head form among the descendants of immigrants; race relations and the pros and cons of assimilation were thereby brought to the fore.

Then, toward the end of the Roaring Twenties, when public concern about Freud, the flapper, flaming youth, bathtub gin, companionate marriage, love in the machine age, and the Babbitt-baiting of Mencken and Sinclair Lewis was at its height, Margaret Mead, one of several able women trained by Boas, began to attract attention. Before long *Coming of Age in Samoa* and *Growing Up in New Guinea* became household handbooks for emancipated youngsters and doting parents, and anthropology emerged from the museum, the study, and the field into the lending library and *The New Yorker*.

The expansion of anthropology continued into the nineteen-thirties, stimulated by generous grants from the Carnegie Corporation, the Social Science Research Council, and similar agencies, in the belief that it was the ideal connective medium for interdisciplinary research enterprises. Anthropology went far to justify that belief; books such as Ralph Linton's *The Study of Man* (1936) showed how the walls between social psychology, sociology, and ethnology could be effectively pierced. To do this Linton took the old Maine and Spencer concept of status, related it to George Herbert Mead's concept of role, distinguished between ascribed and achieved status, and developed, both implicitly and explicitly, a concept of social structure that was more than abstractly schematic. Later he helped to provide, following lines similar to those already established by Hallowell, some rationale for the mushrooming interest in "personality and culture," although that rationale unfortunately failed to impress uncritical enthusiasts such as Margaret Mead, Gregory Bateson, Geoffrey Gorer, and like exponents of diaper determinism. Linton had gone through the Boas mill, but he was also influenced by the sociologist E. A. Ross, and by the psychologist-sociologist Kimball Young, who did so much to disseminate knowledge of the work of George Herbert Mead. Through his effect on Talcott Parsons, several of Linton's concepts, in slightly modified form, bid fair to become part of the standard equipment of American sociologists. Here "who to whom" is not hard to trace.

Linton's 1945 symposium, *The Science of Man in the World Crisis,* did a good deal to generate interest in anthropology as inclusive of all the social science disciplines, the humanities, and those natural sciences having to do with *Homo sapiens.* What will be the outcome of this interest, which is today quite strong, is not easy to forecast. The present symposium may provide some clues. Sociologists, some of whom in times past have made claims quite as extensive, are not kindly disposed toward anthropology as *"the* science of man" —as might be expected.[53]

The expansion of anthropology in "problem" directions to which reference has already been made continued, after a slight lull in the middle nineteen-thirties, to bring it into new fields. For example, the versatile Margaret Mead conducted several researches purporting to demonstrate that the supposedly universal masculine and feminine traits are wholly determined by culture; she produced evidence

---

[53] Here may be instanced articles by Adolphe Tomars, "Some Problems in the Sociologist's Use of Anthropology," *American Sociological Review,* 8, 6, Dec., 1943, pp. 625–634, and Robert Bierstedt, "The Limitations of Anthropological Methods in Sociology," *American Journal of Sociology,* 54, 1, July, 1948, pp. 22–30.

to show, among other things, that in certain societies the adult males are pampered "clinging vines," dependent on the psychically sturdier females for livelihood and guidance. In addition to such studies, finally incorporated in *Sex and Temperament in Three Savage Societies,* she also produced a popular book, *Male and Female.* During World War II she undertook the challenging task of interpreting the Americans to the British, and *vice versa.* Recently her timely interest in national character, coupled with her freedom from linguistic embarrassment, has enabled her to make striking pronouncements about the Germans and the Russians, especially the latter.[54] Her influence on textbook sociology has been quite marked; not only the usual general introductory texts, but also those dealing with marriage and the family, social disorganization, and similar matters refer extensively to her writings, especially in the opening chapters.[55] As can be readily inferred, her popular reputation is a wide one, although recently an anthropological rival has arisen in the person of H. Ashley Montagu; he too has a "problem" orientation.

This orientation is also apparent in the anthropological study of work relations and small work groups with which the names of Arensberg, Chapple, Goldschmidt, and Hart have recently been associated. Getting under way in conjunction with the researches of Elton Mayo, such study—also called industrial sociology, factory sociology, and the like—was for a time almost the whole content ordinarily associated with "applied anthropology."

This term, interestingly enough, began to come into use about a decade after the last sociological publication calling itself "applied," the *Journal of Applied Sociology,* changed its name to *Sociology and Social Research.* During the nineteen-forties a Society for Applied Anthropology was founded, and the movement has a journal now calling itself *Human Organization.* The more inclusive title seems warranted, for anthropologists have become of much importance in some kinds of foreign service training, and others are prominent as field technicians in Point Four programs and a host of similar activities far transcending the limits of a generation ago.

These limits have also been transcended, be it noted, in other than the "applied" fields, for instance, the work of W. Lloyd Warner.

[54] Cf. her informative article in A. L. Kroeber, ed., *Anthropology Today,* Chicago, 1953. The bibliography is especially enlightening.

[55] It is not much to the credit of the writers of certain sociological textbooks that the anthropology they insert frequently has little systematic relation to the rest of the book.

In the nineteen-thirties Warner launched an elaborate many-volumed community study of "Yankee City" that was also an analysis of a New England class system, and the subsequent impact, favorable or unfavorable, on sociologists dealing with social stratification, education, and youth affairs has been very definite indeed. Even though it may be granted that the sociologist O. C. Cox, in *Caste, Class, and Race* (1948), did much to call the accuracy of Warner's methods into serious question,[56] it is still true that the latter is the most influential writer on social stratification in the United States. Previous to his work, few if any anthropologists had ever devoted any particular attention to stratification in the modern societies of the Western world, much less exerted any influence on sociologists dealing with such topics.

The impact of Warner's work on community studies was also well marked; the younger American anthropologists soon developed much interest in villages, small towns, "factory farms," and so on. Some time previous to this, however, the sociologist Robert E. Park, among others, had communicated a strong interest in community studies to his anthropologist son-in-law, Robert Redfield. The kind of community in which Redfield became interested, to be sure, was what at about the same time had begun to be called "folk" by the sociologist Howard W. Odum, but in Redfield's case, naturally enough, the connections with traditional anthropological studies were far more definite. However, Redfield made his "folk society" a polar opposite of urban society,[57] and it seems clear that Park's urban interest played some part in this. But in any case, the direct stimulus to make the community studies that have recently become so numerous came in far greater measure from the anthropologists than the sociologists, even for the rural United States, where one might suppose that the rural sociologists would have a near-monopoly. Linton, for example, did much to interest James West (pseud.) in the "Border South" small town depicted in *Plainville, U.S.A.*, and John Gillin, who received some of his training with Linton, has launched a series of community studies of which the first, *Plantation County*, by Morton Rubin, deals in an excellent way with a rural locale in the Deep South. Even the trail-breaking *Middletown* and *Middletown*

---

[56] Although he also aimed at several others, including Gardner and Dollard.

[57] These categories have been subject to much criticism; the *Through Values* volume mentioned on p. 110 summarizes a good deal of it on pages 45, 254–257, *et passim* (use index). See also George M. Foster, "What Is Folk Culture?" *American Anthropologist*, 55, 2, Part 1, Apr.–June, 1953, pp. 159–173.

in Transition, written by the sociologists Robert and Helen Merrill Lynd, provide only partial exceptions to the increasingly general rule of anthropological influence, for the authors prominently acknowledged their debt to the anthropologist Clark Wissler, whose *Man and Culture* (1923) had wide circulation among social scientists of every description.

The recent contraction of American sociology, in part self-initiated, in part by default, must be placed in what the present writer feels to be regrettable contrast with this expansion of anthropology. It is one thing to forswear academic imperialism—and rightly—with statements such as "sociology is here regarded neither as the mistress nor as the handmaid of the other social sciences, but as their sister";[58] it is quite another to become so narrowly specialized, esoterically technical, methodologically hypnotized, and autistically gobbledy-gookish that reasonably intelligible general sociology at times seems to be on the verge of disappearing altogether.

A trend having like consequences is apparent even where the range of evidence utilized by the American sociologist is concerned. William Graham Sumner, for his *Folkways,* combed the ethnographic and historical literature of the world in more than a half-dozen languages; he practiced sociological research in the grand manner. His method was at times crudely illustrative rather than genuinely comparative,[59] but he assiduously searched for negative instances along with the positive; his mind was by no means of the thesis-defending, single-track type. His follower, Albert Galloway Keller, possessed a less judicial temperament; nevertheless, the massive Sumner-Keller *Science of Society* is worthy of respectful attention, for out of it grew the classificatory system on which are based the files of the amazingly comprehensive Yale Cross-Cultural Survey (now called Human Relations Area Files, Inc.). Without the facilities for ready comparison of key features of several hundred societies scattered over the entire globe, such contributions as Murdock's recent *Social Structure* would have been quite impossible.

Murdock once regarded himself as primarily a sociologist, and the present writer still so regards him, but American sociologists have slowly diminished the cross-cultural scope of their researches while

[58] This appeared in 1932 on p. vii of the *Systematic Sociology* cited on p. 131.

[59] See *Through Values* (cited on p. 110), pp. 132, 141–188, 217–218, and especially 241–250.

the anthropologists have been quite properly maintaining and expanding theirs. Today Murdock calls himself an anthropologist. Everyone takes for granted the cross-cultural or comparative interest of the anthropologist, whereas an increasing number of sociologists are finding it necessary to call themselves "comparative sociologists" (although not necessarily in the Radcliffe-Brownian sense only) if they are not to be confused with those who think it quite proper to be almost wholly ignorant of ethnography, history, and contemporary civilizations other than their own, as long as they have painstakingly counted the number of parking meters in Punkin Center on July 17, 1953.

Even if one were to grant that valid comparison is at times difficult when the classificatory scheme too finely divides the societies surveyed, it is still true that firsthand descriptive studies of ongoing social units are not today made by any significant proportion of American sociologists as sociologists. Some "survey research" occasionally provides relatively full coverage of complete societal groups, but this is more often than not a fortunate accident, given the increasingly specialized aims of such research.

Indeed, if methodology, as the body of basic mathematical, logical, and similar considerations peculiar to a particular discipline, is separated from special techniques such as rating scales, spot-maps, questionnaires, graphs, tables, formulas, and so on, it can be said that sociology is becoming severely limited by its own elaborate skills. Methodology is slighted, but technology, at times almost verging on gadgetry, is having a tremendous vogue. Direct experience in the gathering of primary-source evidence is thought to be unnecessary; such "merely descriptive" effort, as distinguished from "rigorous analysis," can be assigned to assistants. The present writer vividly recalls what Robert E. Park once said about such matters to a group of graduate students.

You have been told to go grubbing in the library, thereby accumulating a mass of notes and a liberal coating of grime. You have been told to choose problems wherever you can find musty stacks of routine records based on trivial schedules prepared by tired bureaucrats and filled out by reluctant applicants for aid or fussy do-gooders or indifferent clerks. This is called "getting your hands dirty in real research." Those who thus counsel you are wise and honorable; the reasons they offer are of great value. But one thing more is needful: firsthand observation. Go and sit in the lounges of the luxury hotels and on the doorsteps of the flophouses; sit on the Gold

Coast settees and on the slum shakedowns; sit in Orchestra Hall and in the Star and Garter Burlesk. In short, gentlemen, go get the seat of your pants dirty in real research.[60]

Anthropology has thus far been saved from being swamped by procedures akin to those now threatening to overwhelm the sociologist only because of its stress on "real research" stemming from its long and sound field work tradition. That tradition now needs much more explicit statement and systematic communication if its advantages are to be conserved.

The recent contraction of the range of sociological interest is also manifest in the acceptance of psychological restrictions in the study of social psychology. Instead of observing man in his ordinary habitat, as it were, *à la* Carpenter's field studies of the gibbon, he is studied in the laboratory cage, with all the sociological drawbacks of such convenient and for some purposes advantageous practice. It is of course very much worthwhile to learn how man behaves in cage situations, but it is risky to generalize from these to unhampered everyday life in the notoriously wide range of cultures that he has produced. Field-observational social psychology has nevertheless lost ground to the cage-study type, largely because the latter offers facilities for what are assumed to be precise and replicable experiments. Now, the sound-proof room, the one-way screen, the sound camera, and well-trained subjects undeniably increase a certain kind of accuracy within the narrow limits they impose. Whether such limits unavoidably diminish accuracy of other kinds is the sixty-four dollar question; perhaps the winning answer is the usual "It depends on the problem." The anthropologists seem to be most interested in the problems of man outside the cage.

Small-group research in many respects necessarily transcends the limits of social psychology, but here too the sociologists are accepting psychological restrictions. For all his vagaries, the sociometrist J. L. Moreno has always stressed the unrehearsed and spontaneous, herein remaining within the Simmel-Wiese sociological tradition, but today many small-group researchers are operating within rigid psychological laboratory limits, utilizing the somewhat elaborate spatial and "field" analogies of Kurt Lewin and J. F. Brown. It is to the credit of anthropologists such as Chapple and others, whose studies of small groups have already been mentioned, that they have relied primarily on "work-habitat" observation. Even the use of the inter-

[60] From brief notes taken by the present writer in April, 1928.

action chronograph has not, among the anthropologists, brought with it the confinement of the laboratory.

It seems fair to say that only the sound emphasis on firsthand observation of unstaged conduct, in unhampered situations, and, on occasion, with proper cross-cultural comparison—in short, only the equivalent of good anthropological field work—will make small-group studies of any far-reaching sociological significance. Otherwise they will remain as sadly limited in the applicability of their conclusions as they are at present.

But it is high time that invidious comparisons cease and constructive comments be offered! Mention has already been made of the community studies recently carried out by anthropologists and those proximately influenced by them; such work with reasonably well-demarcated, coherent, and self-sustaining societal clusters can be of great service in rendering sociological analysis more relevant and less esoteric. The concepts of structure and function, for example, can then be more readily viewed for what they are worth—and when kept in constant interaction with the empirical evidence, they may be worth a good deal.

The same is true of the anthropological studies of "personality and culture." Although it may be granted that much nonsense has been perpetrated by some exponents of this trend, it is also true that it seems to offer fruitful leads not now properly followed up by social psychologists and sociologists. They tend to be properly critical of "intuitionism," "holism," and various brands of variously orthodox psychoanalytic dogma purveyed by certain anthropologists, but they do not offer constructive suggestions or positively oriented research. In combination with the contributions of French, German, and British social scientists mentioned elsewhere in this and other chapters, it should be possible, with appropriate care, to extract much sociological profit from the daring dashes of American anthropologists into the "personality and culture" wilderness.

And, at the risk of repetition, the sociologists can learn a good deal from the anthropologists by dealing with culture a little more systematically and comprehensively, rather than as the occasional vehicle of ritual invocations in the opening chapters of elementary textbooks. Parsons has clearly profited from his collaboration with anthropologists in the Department of Social Relations at Harvard; his stress on culture as the context of social action provides firmer timbers for an interdisciplinary bridge that was previously even more frail and hard to cross. Similarly, other sociologists have been aided by the anthropological emphasis on values and value-systems as the absolutely essential aspects of culture; some of them had long been fairly well aware of this because of the examples of Sumner and of Thomas and Znaniecki, but in relation to the present sociological generation in the United States the anthropologists have been and perhaps for some time will continue to be the most useful guides.

VI

However, all of these suggestions, leads, contributions, and so on are rather scattered; intellectual staging centers for American interdisciplinary expeditions are needed. What, in the present and the near future, seems to be the region where the most effective collaboration of "the social sciences," or "the behavioral sciences," or "the sciences of action," or "the sciences of social relations," or "the sciences of social man" is likely to take place?

For the present writer, at least, that region is constituted by what has been somewhat inaptly called "area research" or "area studies"—but the name need not worry us. Let us look at an example: the institutes recently established in many American universities for instruction and research on Russian affairs pool the resources of all the social sciences, the humanities, and some of the natural sciences insofar as they have any bearing on the problem of how the peoples of the "Russian area" come to behave as they do, and what can be predicted and done about it. While recognizing the absolutely indispensable contributions to such "area studies" that have been and are being made by human geography, economics of several varieties (e.g., land economics), political science, history, and so on, it still seems true that anthropology and sociology, together with social psychology, fill key organizing roles.[61]

It is freely granted that area studies are not pure science, and that if viewed as other than means of eventually applying abstract principles to concrete situations, and then deriving corrections and fresh applications of those principles, they can be so restrictive as to be seriously damaging. Moreover, teams of specialists do not achieve worthwhile scientific results merely by saying, "Go to, let us cooperate." Good will must be translated into advanced and at times

[61] See Julian H. Steward, *Area Research: Theory and Practice,* New York: Social Science Research Council, 1950.

arduous education on the job; every specialist must learn the essentials of the work of every other specialist, and this is not done with mirrors. Anyone who believes that area research that merits the name of science achieves integration by binding together in one volume the self-sufficient reports of isolated savants is in grievous error. *Integration must be achieved in the minds of the collaborators;* every section of the finished project, whatever the emphasis required by the particular occasion for the research, must reflect the relevant aspects of every other section.[62]

There are of course many ways in which a workable scientific division of labor can be set up in area research, and to discuss them for every discipline would exceed both the present writer's competence and these space limits. Comment will be restricted, then, to anthropology and sociology, and confined to the more abstract features.

Every area research must do justice to the unique characteristics of the body of phenomena being studied. Description of the relatively unique, or idiography, provides the anthropologist (of whom it can usually be assumed that he has had appropriate training in firsthand and close-range observation) with the opportunity to provide a firm empirical base for whatever conclusions may be reached.

Such conclusions necessarily will have some degree of generality if they are to be in any respect scientific; that is, they will be nomothetic. Here the sociologist, with his long albeit neglected comparative tradition, and the anthropologist, sensitized to the unique but with comparative work part of his daily routine in many instances, can make sure that the basic requirements for valid generalization are adequately met.[63]

Where the procedures of area research are concerned, these can be worked out by the anthropologist and the sociologist together, in conjunction with available philosophers of science, logicians, mathematicians, and statisticians thoroughly familiar with the "area universe" in question. Such procedures comprise both (1) the fundamental methodological considerations of anthropology and sociology in their various subdivisions, together with their collaborative disciplines, and (2) the specific techniques necessary to bring the methodologies into working relation with the empirical evidence.

---

[62] G. P. Murdock, "The Conceptual Basis of Area Research," *World Politics,* 2, 1951, pp. 571–578.

[63] For reasonably full discussion of the idiographic and the nomothetic, see *Through Values,* cited on p. 110, using index and section bibliographies.

And throughout area research, as in every other research of truly scientific rather than speculative character, the informing spirit must be like that thus expressed by Ratzel:

No matter how important the tools of methodology and classification may be, we shall always stray into dilettantism if we make them alone, apart from the testing power of monographic research, the object of scientific thought. . . . Dilettantism, indeed, is constituted more by illusion about depth of problems than by ignorance of methods, and naive optimism about this depth is therefore most characteristic of the dilettante. In the long run the frontiers of science are never defined by methodological claims alone, for these bear the same relation to creative research as the mere rehashing of books does to the hand-to-hand struggle with Nature for her secrets.[64]

If we can get convergences among the sciences of social man in the realm of research rather than merely in the pages of a symposium such as this, there will be cause for jubilation. In the meantime, let each one of us do his duty as he sees it, properly considerate of the claims of others, but never becoming so apologetic about his own that he merits the attentions of Maud the Mule.

[64] Friedrich Ratzel, *Anthropogeographie*, Stuttgart, 1882–1889, Part II, p. xi, trans. the present writer's.

CHAPTER 6

Psychology and Anthropology

A. IRVING HALLOWELL

I. Introduction

IF THE RELATIONS between anthropology and psychology[1] are considered merely abstractly or without awareness of historical circumstances, any appraisal of the present situation or their future relations will lack clarity and depth. Although there have been discussions, particularly in the past three decades, of the more general, as well as certain specific, aspects of these relations, and research in

[1] For the purpose of this chapter *psychology* is defined in the broadest terms; so is *anthropology*. In effect, the nature and range of psychology coincides with the conception of this discipline as set forth in *The Place of Psychology in an Ideal University*, Cambridge: Harvard University Press, 1947, p. 2; the Report of the University Commission to Advise on the Future of Psychology at Harvard, Alan Gregg, chairman, C. I. Barnard, D. W. Bronk, L. Carmichael, J. Dollard, T. M. French, E. R. Hilgard, W. S. Hunter, E. L. Thorndike, L. L. Thurstone, J. C. Whitehorn, R. M. Yerkes:
"Psychology is actually what psychologists do and teach: defined briefly, it is the science of human and animal behavior, both individual and social. To expand this definition, psychology is the systematic study, by any and all applicable and fruitful methods, of organisms in relation to their behavior, environmental relations, and experience. Its purpose is to discover facts, principles, and generalizations which shall increase man's knowledge, understanding, predictive insight, directive wisdom and control of the natural phenomena of behavior and experience, and of himself and the social groups and institutions in which and through which he functions. Psychologists seek to provide a basic science of human thinking, character, skill, learning, motives, conduct, etc. which will serve all the sciences of man (e.g., anthropology, sociology, economics, government, education, medicine, etc.) in much the same way and to the same extent that biology now serves the agricultural and medical sciences."

some areas of common interest has taken place,[2] no one has yet writ-

[2] See Alfred C. Haddon, *History of Anthropology*, 1st edition, New York: Putnam, 1910; 2nd edition, London: Watts and Company, 1934; T. K. Penniman, *A Hundred Years of Anthropology*, New York: Macmillan, 1936; Robert H. Lowie, *The History of Ethnological Theory*, New York: Farrar and Rinehardt, 1937; Wilhelm Mühlmann, *Methodik der Völkerkunde*, Stuttgart: Enke, 1938.

More than a quarter of a century ago A. A. Goldenweiser, writing a chapter on "Cultural Anthropology" that appeared in H. E. Barnes, *The History and Prospects of the Social Sciences* (1925), made the interesting observation that:

"In reviewing the work of Rivers and the American anthropologists I took occasion to note a revival of interest in psychological problems which arose in the course of cultural studies. The American students, in particular, were keenly conscious of the necessity of a psychological technique to supplement the objective studies of the historian of culture. But to recognize this necessity was one thing, to supply the technique another. The attempts in this direction first of all produced a crop of theoretical discussions dealing with the general relations of psychology and sociology, discussions which bring to mind the old controversy over the folk-soul between Wundt on the one hand and Steinthal and Lazarus on the other, as well as the prolonged disquisitions among sociologists over the content and nature of sociology."

The references Goldenweiser gives are as follows: Robert H. Lowie, "Psychology and Sociology," *American Journal of Sociology*, Vol. 21, 1915, pp. 217–229; W. H. R. Rivers, "Sociology and Psychology," *Sociological Review*, Vol. 9, 1916, pp. 1–13; A. L. Kroeber, "The Possibility of a Social Psychology," *American Journal of Sociology*, Vol. 23, 1918, pp. 633–651; A. M. Hocart, "Ethnology and Psychology," *Folk Lore*, Vol. 75, 1915, pp. 115–138; R. R. Marett, *Psychology and Folk Lore.*

Goldenweiser himself had by this time published "Psychology and Culture," *Publications of the American Sociological Society*, Vol. 19, 1925, pp. 15–24 and, in 1927, the chapter he contributed to *The Social Sciences and Their Interrelations*, W. F. Ogburn and A. A. Goldenweiser, eds., was entitled "Anthropology and Psychology." Both of these were later reprinted in the volume of his selected papers, *History, Psychology, and Culture*, New York: Knopf, 1933. No reference was made by Goldenweiser to the significant article published in 1924 by Marcel Mauss, "Rapports réels et pratiques de la psychologie et de la sociologie," *Journal de Psychologie Normale et Pathologique*, nor to the Presidential Address of C. G. Seligman, "Anthropology and Psychology: A Study of Some Points of Contact," *Journal of the Royal Anthropological Institute*, Vol. 54, 1924.

In present-day perspective, what is significant about all of the citations made by Goldenweiser is that they antedate the subsequent literature of the personality and culture movement, so that not one of them, including Goldenweiser's own articles, appears in Haring's bibliography, *Personal Character and Cultural Milieu: A Collection of Readings*, compiled by Douglas G. Haring, revised ed., Syracuse University Press, 1948. Of the additional items to which I have called attention Seligman is given, but Mauss is not. Although Goldenweiser contributed a chapter entitled "Some Contributions of Psychoanalysis to the Interpretation of Social Facts" to *Contemporary Social Theory*, edited by H. E. Barnes, Howard Becker, and Frances B. Becker, he is concerned with various psychoanalytic writers and their critics rather than with the research, already begun, in the personality and culture area.

In 1939 Gillin published a review article, with bibliography, of the literature dealing with "Personality in Preliterate Societies," *American Sociological*

ten a detailed history of all the varied influences and counterinfluences exerted at successive periods in the development and expansion of these disciplines. Nevertheless, it seems desirable to survey briefly some of the relevant facts even if our historical résumé must be sketchy and incomplete.

The most outstanding impression received when this over-all viewpoint is adopted is that there have been many more significant connections than are ordinarily supposed, at least from the anthropological side, even if these have not always led to cooperative research; some have even provoked negative reactions. Anthropologists generally have been far from insensitive to psychology. Thus, while no exception need be taken to Kluckhohn's statement that "the dominant currents in American anthropology prior to, roughly, 1920 were descriptive and historical," and "awareness of *psychiatry* in other than the most diffuse sense that there was such a thing as 'mental disease' was almost completely lacking," it is misleading to say that "the prevalent trend of American anthropology was *anti-psychological*" (italics mine).[3] If psychology is given its broadest connotation and anthropology considered in all its historical aspects, it is apparent, on the contrary, that from its beginnings, both abroad and in America, anthropology has been almost continuously influenced by psychology, but not always with respect to the same problems. Writing in England Haddon seems to have been aware of this. His *History* contains a brief chapter entitled "Comparative Psychology" in the first edition (1910), and an expanded chapter "Individual and Ethnic Psychology" in the revised edition (1934). Referring to this country he says, "In the United States of America, thanks to the influence of Boas, psychology has been well recognized by students as an essential factor in ethnology."[4] It may likewise be noted in passing that the founder of the Bureau of American Ethnol-

Review, Vol. 4, pp. 681–702; and in 1944 Kluckhohn systematically analyzed the concrete contributions which reflected "The Influence of Psychiatry on Anthropology in America during the Past One Hundred Years," *One Hundred Years of American Psychiatry*, edited by J. K. Hall, G. Zilboorg, and H. A. Bunker, New York: Columbia University Press. For additional guidance to the personality and culture literature see Haring, *op. cit.*; *Personality in Nature, Society, and Culture*, Clyde Kluckhohn and Henry A. Murray, eds., 2nd edition, 1953, New York: Knopf; and A. Irving Hallowell, "Culture, Personality, and Society" in *Anthropology Today*, A. L. Kroeber, ed., Chicago: University of Chicago Press, 1953.

[3] Kluckhohn, 1944, *op. cit.*

[4] Haddon, *op. cit.*, p. 68. This statement only occurs in the revised edition.

ogy, J. W. Powell, in an article on "Anthropology" published in the *Universal Cyclopaedia* (1900), discussed the subject under the headings: Somatology, Psychology, Ethnology.[5]

It is only if psychology is defined very narrowly, or the influence of closely related specialties like psychiatry or psychoanalysis is considered, or when the traditional interests of specialized groups of anthropologists, such as linguists or archeologists, are referred to, that a complete absence of interest or influence before 1920 can be alleged. For nearly a century, many outstanding anthropologists have been interested in various psychological problems; and it has always been quite generally recognized that such problems were intimately connected with some of the large, central, and perennial questions that anthropologists sought to answer. Besides his concern with *Elementargedanken* and *Völkergedanken*, Bastian, for example, was interested in abnormal psychology; he reported firsthand observations on possession in a psychological journal.[6] Among other things, Waitz posed the question of national character: "The question has been frequently asked," he says, "in what consists the national character of a people? The preceding investigation has shown that it depends on so many conditions that an exact analysis is extremely difficult. That it is not the race alone which determines it, is proved by there being different nationalities within the same race. It is therefore probable . . . that the mental peculiarities of peoples are generally more flexible and changeable than the physical characters of the race, and are transmitted with a less degree of constancy."[7] Andrew

[5] The special interest of this article does not lie in its concrete content, but in the fact that Powell wrote it and that he refers to psychology as "an integral part of anthropology." He finds it necessary to distinguish two basic schools of psychological thought: one that claims psychology as a natural science, and a "second school, which is earlier in history, and still the larger," which assumes the existence of a "spiritual soul" and has intimate connections with philosophy.

[6] P. W. A. Bastian, "Ueber psychische Beobachtungen Naturvölkern," Publications of *Die Gesellschaft für Experimental-Psychologie zu Berlin*, Leipzig, 1890, Vol. 2, pp. 6–9. T. K. Oesterreich, *Possession: Demoniacal and Other Among Primitive Races, in Antiquity, the Middle Ages, and Modern Times*, New York: Richard R. Smith, 1930, p. 378 says: "The extraordinary importance accruing to the phenomena of possession amongst primitive races has hitherto been insufficiently appreciated by ethnology. One single ethnologist, Adolf Bastian, whose numerous works have not attracted the attention they deserved owing to their abstruse literary form, was fully alive to it. In his works we meet possession at every turn, and their unsupported testimony would be adequate to demonstrate its significance in the savage world."

[7] Theodore Waitz, *Introduction to Anthropology*, London, 1863, edited by J. Frederick Collingwood with numerous additions by the author, from the first volume of *Anthropologie der Naturvölker*.

Lang was interested in parapsychology, and declared that "anthropology must remain incomplete while it neglects this field, whether among wild or civilized men."[8] Frazer explicitly articulated his concern with mental as well as cultural evolution.[9]

Whatever the professional preoccupations of anthropologists have been, one inescapable problem has always directed their attention to psychology: some basic assumptions always had to be made about the psychological nature of man and human behavior. These assumptions had either to be derived from "common sense" or some attention paid to the data and conclusions of psychology as a recognized discipline, irrespective of the latter's stage of development or the specialized preoccupations of the psychologists of any particular period. For instance, the question of racial differences in "mentality," especially intellectual superiority or inferiority, involved a problem central to all anthropological research. Could a basic "psychic unity" of mankind be assumed, or was it necessary to consider the possibility that there were innate racial barriers to "culture progress" or even to the acquisition of "higher" cultural forms by the "lower" races in a process of "diffusion"? Boas, writing in 1910, said that "the ultimate aim of Waitz's great work is the inquiry into the question whether there are any fundamental differences between the mental make-up of mankind the world over, racially as well as socially."[10] But in Waitz's time, and for a long period thereafter, it was not possible to approach this question by means of psychological tests of any kind. Yet the assumption of psychic unity became traditional, nonetheless.

Furthermore, as more and more data were accumulated and characteristic problems defined, it is not surprising that the attitudes of anthropologists toward psychological concepts and formulations

[8] Quoted by Haddon, *op. cit.*, p. 61, with reference to *The Making of Religion*, 1898. In his *Cock Lane and Common Sense* (1894) Lang had first called attention to "savage spiritualism," which he said "wonderfully resembles, even in minute details, that of modern mediums and séances. . . ." A few contemporary anthropologists have manifested interest in parapsychology. See, e.g., Victor Barnouw, "Siberian Shamanism and Western Spiritualism," *Journal of the American Society for Psychical Research*, Vol. 36, 1942, pp. 140–168; and "Paranormal Phenomena and Culture," *ibid.*, Vol. 40, 1946, pp. 2–21. Margaret Mead has served as a member of the Board of Trustees of the American Society for Psychical Research and, in March 1953, John R. Swanton in a communication directed to "Fellow Anthropologists," drew attention to this field.

[9] James G. Frazer, "Scope and Method of Mental and Anthropological Science," *Science Progress*, Vol. 16, 1922.

[10] Franz Boas, "Psychological Problems in Anthropology," *American Journal of Psychology*, Vol. 21, 1910, pp. 371–384.

should increasingly reflect their own mastery of data, special interests, and conclusions. The negative reaction to intelligence tests as a reliable means of determining innate mental differences attributable to race was not only an indication of a continuing interest in the question of the psychic unity of man; it was a determined effort to demonstrate the direct relevance of the data and concepts of physical and cultural anthropology to a basic psychological problem. The instinct doctrines formulated by McDougall and others became another pertinent focus of anthropological knowledge and appraisal. Thus anthropologists have not hesitated to make independent judgments of what psychologists have had to offer them.

It has now become increasingly apparent, moreover, that psychological problems which were not of special interest to contemporary psychologists, or not even recognized by anthropologists themselves, have nevertheless been inherent in certain data and concepts of anthropology. Tylor's classical definition of culture, for example, asserts that culture is acquired. This simple statement actually embodies a host of psychological problems of no particular interest to the psychologist of Tylor's day. Again, anthropologists have been able to make use of the fact that culture is transmitted in human societies from one generation to another, without exploring the actual psychological processes and consequences involved in it. Similarly, while the empirical facts of diffusion were so ardently studied in detail, the psychological problems inherent in the transmission of culture traits and complexes from one group of people to another were not dealt with at all. The recent interest in personality and culture is by no means a completely unique phenomena: it only represents a more explicit articulation of the kind of psychological problem that is inherent in anthropological data. For the point of departure is the old and well established anthropological dictum that culture is acquired by the individual and that empirically, one culture may vary greatly from another in pattern and content. So the question is raised: What psychological significance have these facts with reference to the differential psychological structuring and behavior of one group of people as compared with another?

Since both psychology and anthropology have grown up as autonomous and rapidly developing disciplines, it was not to be expected that at any particular time the major problems and interests of one would be found immediately useful to the other. In the first place, there has been no established and unchanging body of data, concepts, principles, and laws characterizing psychology as a unified science at any one period. What we have is not only a developing but a rapidly expanding discipline, or congeries of disciplines. Reviewing the preceding five decades of psychology in 1940, Bruner and Allport[11] noted thirty-seven types of psychological interest and distinguished and characterized the shifting foci of these interests from decade to decade. All this is mirrored in the twenty divisions of the American Psychological Association as it exists today.

Although Kroeber has remarked[12] that "psychology is apparently less developed than anthropology," on the contrary, the diverse and expanding interests of anthropology in recent decades parallels the history of psychology. Anthropology is not a tightly knit and unified discipline either,[13] nor is it concerned with problems that have been traditionally defined to the exclusion of others any more than is psychology. The demonstrable fact is that the expanding interests of both disciplines, while formally approaching the study of man from quite different points of view, are more and more focusing on certain common problems and overlapping fields of interest.

What has been happening in psychology is inadvertently specified in a review of the second edition of Boring's *History of Experimental Psychology* (1950).[14] The reviewer says that psychologists "will ask but one question about it: Does it now really go beyond the development of structural psychology; beyond sensation, perception, and the classical treatment of the higher mental processes; beyond that somewhat limited group of men who arrogated to themselves the name of experimental psychologists and, by implication, relegated the student of learning, the mental tester, the social psychologist, and others to the fields only partially within the realm of science? There is no simple answer to the question." Perhaps not. But is it not significant historically that the reviewer also stresses the fact that to the author "psychology's greatest names are Darwin and Freud, Helmholtz and James" and that "Boring is no longer concerned with whether one can properly call these men 'experimental psychologists' or even 'psychologists.'" Besides this, it is precisely

[11] Jerome S. Bruner and Gordon W. Allport, "Fifty Years of Change in American Psychology," *Psychological Bulletin*, Vol. 37, 1940.

[12] A. L. Kroeber, *The Nature of Culture*, copyright 1952 by the University of Chicago, Introduction to section entitled "Psychologically Slanted," p. 299.

[13] See the Wenner-Gren Symposium volumes, *Anthropology Today* (1952) and *An Appraisal of Anthropology Today* (1953).

[14] Charles W. Bray in *Science*, Nov. 3, 1950.

in such areas of psychological investigation as learning, mental testing, and social psychology, that the recent interests of the anthropologist have been found to overlap those of the psychologist.

At earlier periods, however, a comparable situation existed. If we
go back to the first half of the nineteenth century, we find that, in
certain instances, anthropologists and psychologists were not so
sharply distinguished as subsequently and that it was a well recognized fact that they had some common interests. In the second half
of the century, under the impact of the theory of biological evolution in the post-Darwinian period, new areas of this kind emerged. In
the *Descent of Man* (1871) and the *Expression of the Emotions in
Man and Animals* (1872) Darwin himself insisted on psychological as
well as biological continuities between man and other animals so
that these books precipitated even more violent reactions than his
earlier *Origin of Species.* Romanes coined the term "comparative
psychology" and began to write about mental evolution and the evolution of intelligence. (*Mental Evolution in Animals,* 1883.) Once im-
bued with the evolutionary idea, both psychologists and anthropologists envisaged wider and wider areas that seemed to demand integral treatment in a common frame of reference. If evolution was to
be accepted as a process that transcended the biological realm, it
did not stop with the emergence of man. There was a further question of vital importance. Had not mental as well as cultural evolution
taken place in the course of man's long career upon this earth? Sir
James G. Frazer, who is so often cited in anthropological literature
as a typical example of late nineteenth-century evolutionary thinking as applied to human *cultural* development, may, at the same
time, be cited as an example of an anthropologist who defined his
own basic orientation with reference to a psychological problem,
that is, the question of *mental* evolution. Frazer held that not only
ethnographic data, but the ontogenetic development of the child and
the study of patients in mental hospitals could contribute to this
problem. The key to his position is to be found in his acceptance of
the theory of recapitulation.[15] "It is a reasonable inference," he says,
"that, just as the development of their bodies in the womb repro-

---

[15] See A. Irving Hallowell, "The Child, The Savage, and Human Experience," *Proceedings of the Sixth Institute on the Exceptional Child,* Langhorne,
Pa.: The Woods Schools, 1939, pp. 8–34; reprinted in Haring, *op. cit.,* for a
critical appraisal of the recapitulation theory in psychology and psychoanalysis.

duces to some extent the corporeal evolution of their remote ancestors out of lower forms of animal life, so the development of their minds from the first dawn of consciousness in the embryo to the full light of reason in adult life reproduces to some extent the mental evolution of their ancestors in ages far beyond the range of history. This inference is confirmed by the analogy which is often traced between the thought and conduct of children and the conduct of savages; for there are strong grounds for holding that savage modes of thinking and acting closely resemble those of the rude forefathers of the civilized races. Thus a careful study of the growth of intelligence and of the moral sense in children promises to throw much light on the intellectual and moral evolution of the race."[16]

Since Frazer took the position that "the first broad and sharp division" in the general science of man "is between the study of man's body and the study of his mind," he expressed a preference for "the more general name of mental anthropology" for the division of the subject in which he himself worked, rather than the term "social anthropology." Taken at its face value there was no attempt here to exclude psychological questions on any *a priori* grounds from anthropological consideration but rather categorically to embrace them. What Frazer firmly believed was that anthropology had a specific psychological contribution to make to the study of the human mind. To him the province "of mental or social anthropology" was "the study of the mental and social conditions of the various races of mankind, especially of the more primitive races compared to the more advanced, with a view to trace the general evolution of human thought, particularly in its earlier stages. This comparative study of the mind of man is thus analogous to the comparative study of his body which is undertaken by anatomy and physiology."

While it may seem strange to us now to characterize Frazer as a psychologically oriented anthropologist, nevertheless, I believe that it was the particular kind of psychological interest which Frazer represented that explains why Freud turned to him and the use he made of Frazer's work. For Freud, too, was concerned with mental evolution as a generic human problem, as well as with psychological development in the individual. And he, too, embraced the concept of recapitulation and attempted to relate the neurotic and the sav-

[16] James G. Frazer, *op. cit.*, p. 586.

age. Curiously enough, Frazer was antipathetic to psychoanalysis and never read Freud.[17]

The psychological counterpart to the broad-gauged evolutionary approach articulated by Frazer has been variously labelled "comparative," "genetic," or "developmental" psychology.[18] While some psychologists of an older generation, like Hall and Baldwin, adopted the theory of recapitulation and made it an integral part of their thinking, as did Frazer and Freud, this working hypothesis is not a necessary assumption of developmental psychology.[19] Nevertheless, the central importance that was once given to it as well as the fact that, as an auxiliary hypothesis, the inheritance of acquired characters was often invoked as the means whereby the experience of past generations of individuals was assimilated to the biological heritage of succeeding generations, seem to have dampened interest in the genuine problems that exist in this area,[20] for biologists came to re-

[17] B. Malinowski, *A Scientific Theory of Culture and Other Essays*, Chapel Hill: University of North Carolina Press, 1944, p. 182, says that Frazer not only rejected "psychoanalysis and all that it meant. He could never be persuaded to read anything by Freud or his school, in spite of the fact that Freud's anthropological contributions are based on Frazer."

[18] Heinz Werner, *Comparative Psychology of Mental Development*, New York: Harpers, 1940, p. 3, points out that "the concept 'developmental psychology' is perfectly clear if this term is understood to mean a science concerned with the development of mental life and determined by a specific method, i.e., the observation of psychological phenomena from the standpoint of development." On the other hand, "there are certain investigators who, when they use the term 'developmental psychology,' refer solely to the problems of *ontogenesis*. The mental development of the individual is, however, but one theme in genetic psychology. Related to the developmental psychology of the individual is the developmental study of larger social unities, a field of interest intimately linked with anthropology and best known by the name of *ethnopsychology*. The question of the development of the human mentality, if not arbitrarily limited, must lead further to an investigation of the relation of man to animal and, in consequence, to an *animal psychology* oriented according to developmental theory."

[19] *Ibid.*, p. 26.

[20] Wayne Dennis, discussing "Developmental Theories" in *Current Trends in Psychological Theory*, Pittsburgh: University of Pittsburgh Press, 1951, points out that "it would not be inaccurate to say that developmental psychology began with a theory—the theory of recapitulation. Child psychology, the most productive segment of developmental psychology, began shortly after the promulgation of the theory of evolution when all scientific minds were inflamed by this great conceptual achievement." But after the movement had arrived at the "concise hypothesis" that "ontogeny recapitulates phylogeny, . . ." its decline was imminent. The evolutionary viewpoint had seemed to open up wide unconquered scientific vistas to child psychology. But on closer approach these beckoning plains proved to be inhabited only by unsubstantial figures and retreating will-o'-the-wisps. There was not a testable hypothesis in the entire landscape."

ject both the recapitulation theory and the notion that acquired characters could be inherited. Contemporary anthropologists, along with their rejection of unilineal cultural evolution have shied away from developmental psychology. When applied to ethnic groups it immediately recalls the kind of unpalatable formulation of "primitive mentality" expounded by Lévy-Bruhl. The fact remains, however, that developmental psychology is a prime example of the emergence in the post-Darwinian period of a problem area common to anthropology and psychology that poses questions that have by no means been solved.[21]

The relations of anthropologists and psychologists, therefore, should be thought of in terms of the changing and expanding interests of both disciplines, of special areas of inquiry, and concrete problems. Are we now entering a period in which there are potentialities present for new and more systematically organized relations that can be directed toward significant research? Are there bodies of data, concepts, and principles in both disciplines that will enable us to unify our approach to the study of man and human behavior?

II. Historical Retrospect

In any historical review of the pioneers of nineteenth-century anthropology the work of two influential figures, Theodor Waitz (1821–1864) and Adolph Bastian (1826–1905) is never omitted. Both of these men had achieved scholarly maturity in the pre-Darwinian period. The first volume of Waitz's *Anthropologie der Naturvölker* had appeared several months before the *Origin of Species* and Bastian's three-volume work *Der Mensch in der Geschichte* was published in 1860. It is likewise well known that both men were psychologically oriented. What has not always been sufficiently stressed, however, is the more concrete fact that they were influenced by a prominent and original figure in early nineteenth-century psychology. Both Waitz and Bastian were Herbartians.[22]

Since Herbart (1776–1841) "represented a departure from the dominant associationist school, inasmuch as he conceived of the mind throughout in terms of dynamic forces rather than of passive

---

[21] See the remarks of Brewster Smith in this volume.
[22] See Paul Honigsheim, "The Philosophical Background of European Anthropology," *American Anthropologist*, Vol. 44, 1942, p. 380.

mechanisms,"[23] these pioneer anthropologists may be said to have followed a progressive trend of their time. Waitz, in fact, had published books on psychology a decade before the first volume of his *Naturvölker* appeared.[24] And Ribot, in his book on German psychology (1885) remarks upon what is to us a paradox today. He says that the psychological titles of some of Bastian's early books are deceptive. They are much more *anthropological* than psychological![25] On the other hand, Lowie says that Waitz's *Naturvölker* "is largely a treatise on primitive mentality" and, as contrasted with Klemm, "the depth of his psychological insight" is noteworthy.[26] While it is ordinarily pointed out that both Waitz and Bastian espoused the notion of the psychic unity of man, Lowie does not refer to Herbart's influence on them. This might be of little interest, were it not for the fact that Lazarus and Steinthal, who first promoted *Völkerpsychologie,* were likewise Herbartians.[27] Quite aside from any evaluation of their achievement, it was from Herbart that they adopted what was at that time a novel idea: "that psychology remains incomplete as long as it considers man only as an isolated individual."[28] Consequently, when Lowie directs our attention to the fact that "decades

[23] J. C. Flugel, *A Hundred Years of Psychology, 1833–1933,* New York: Macmillan, 1933, p. 23. This author finds in Herbart's psychological system certain faint resemblances to Freud. Thomas Ribot, *German Psychology Today,* New York: Scribners, 1886, p. 24, points out that "the first efforts toward a scientific psychology in Germany are due to Herbart. They constitute a transition from the pure speculation of Fichte and Hegel to the unmetaphysical psychology."

[24] Ribot, *op. cit.,* refers to his distinguished place in German psychology.

[25] *Ibid.,* p. 67. One example cited is *Beitrage zur vergleichenden Psychologie. Die Seele und ihre Erscheinungsformen in der Ethnographie,* Berlin, 1868.

[26] Lowie, 1937, p. 16.

[27] Ribot, *op. cit.,* pp. 60 ff.; Fay B. Karpf, in *American Social Psychology: Its Origins, Development, and European Background,* New York: McGraw-Hill, 1932, pp. 42 ff., discusses this movement in relation to Herbart. "Ethnic" rather than "folk" psychology is the preferable English rendering. (See Howard Becker in this volume.)

[28] Ribot, *op. cit.,* devotes a chapter to the "School of Herbart and the Ethnographic Psychology." He says (p. 51): "At first sight it seems strange enough that so concrete a form of psychology should attach itself to the school of Herbart; but, in fact, the disciples have only developed some of their master's views. This point deserves attention, for one would hardly suppose that the founder of the mathematical psychology would have attached great importance to such investigations. He maintains, however, that psychology remains incomplete as long as it considers man only as an isolated individual. He was convinced that society was a living and organic whole, ruled by psychological laws that are peculiar to it." Explicit references to the sources of these statements in Herbart's work are given.

before Rivers [Bastian] argued that a science of mental life must take cognizance of ethnographic data, because the 'individual's thinking is made possible only by his functioning in a social group,' "[29] the psychological source of Bastian's view is obscured if his link with Herbart is not made explicit.

The full implication and expansion of Herbart's germinal idea belongs to the history of social psychology. In the immediately subsequent period psychologists concentrated more and more upon the study of "mind" in the laboratory, rather than in "society,"[30] and

[29] Lowie, 1937, p. 36; see Mühlman, *op. cit.*, for a more extended discussion of Bastian's sociopsychological interests.

[30] It is particularly interesting, therefore, to note that Ribot, *op. cit.*, in the chapter devoted to a discussion of Lazarus and Steinthal, Waitz, and, briefly, Bastian, refers to the famous passage in Hume in which this eighteenth-century philosopher says:

"Would you know the sentiments, inclinations and course of life of the Greeks and Romans? Study well the temper and actions of the French and English. . . . Mankind are so much the same in all times and places that history informs us of nothing new or strange in this particular. Its chief use is ever to discover the constant and universal principles in human nature. . . ."

Ribot (p. 52) then goes on to say:

"In our day we think differently: We believe that this abstract study, amounting to some general characteristics, gives a knowledge of man but not of men: We believe that all who share our common humanity were not cast in a common mould, and we are curious about the smallest of these differences. Hence a new conception in psychology.

"As long as naturalists confined themselves to a pure description of races and species considered as permanent; as long as historians, indifferent to the variations of the human soul in the lapse of ages, spread upon all their recitals the same uniform and monotonous varnish; an abstract psychology, like that of Spinoza and Condillac, seemed the only psychology possible. Nothing else was thought of, and when a very refined and subtle spirit was subjected to minute analysis, it was said of this psychology: It has given us to know man.

"But when the idea of evolution was introduced into the sciences of life and into historical study, stirring and reviewing the whole, psychology felt the impulse. The question was raised: Is this abstract study of man sufficient? Does it give more than broad traits and general conditions; to be simple and exact, does it not need completion? The lower forms of humanity have exhibited particular modes of feeling and action, and the history of civilized peoples has shown variations in sentiment, in social ideals, in moral or religious conceptions, and in the languages that express them. Psychology has profited by it. It occupies, in fact, in the structure of human knowledge a very exact place between biology below and history above."

Ribot here indicates, in principle, his appreciation of the need for a wider frame of reference in the study of human psychology, one which includes cultural variables, in the contemporary sense. But the particular orientation and methodology of Lazarus and Steinthal was not adequate for this task and "orthodox" psychology maintained its basic biological orientation for a long time to come.

anthropologists became more and more concerned with the broad sweep of human history, interpreted in terms of unilinear series of cultural events. Psychology began to attain its independence; the experimental laboratory became the symbol of its scientific status; Wundt was the officiating priest. The problems centered around psycho-physics, visual and auditory sensations, reaction time, and so on. Attention was centered upon responses that could be empirically established as *common* to all subjects, individual differences not being considered significant. It was on the basis of observations of this nature that a science of psychology was to be built up, so that Boring in the first edition of his *History of Experimental Psychology* could define it as "the psychology of the generalized, human, normal, adult mind as revealed in the psychological laboratory."[31]

It was to men belonging to this tradition in psychology that Haddon turned when he organized the Torres Straits Expedition (1898). W. H. R. Rivers had studied in Germany and had begun lecturing at Cambridge on the physiology of the sense organs in 1893. He planned one of the earliest systematic courses in experimental psychology anywhere in the world and the first in England. In 1897 Rivers was appointed Lecturer in Experimental Psychology and the Physiology of the Special Senses.[32] Three others on this expedition famous in anthropological history were C. S. Myers and William McDougall, students of Rivers, who later attained eminence in psychology, and C. G. Seligman, a young physician, who subsequently became a distinguished figure in anthropology. Although the latter's chief contribution to the work of the expedition was made as a medical man, Meyer Fortes says that Seligman "was deeply interested in the ethnological and psychological investigations as well, and took a share in them . . . and saw at first hand the importance of studying the psychology of primitive peoples." Twenty years later, stimulated by his experience as a staff member of the shell-shock hospital at Maghull during World War I and his association there

[31] E. G. Boring, *A History of Experimental Psychology*, New York: Century, 1929, p. viii.

[32] Rivers contributed a chapter on "Vision" to the outstanding *Textbook of Physiology* edited by Sir Edward S. Schafer (1900). Along with James Sully, A. F. Shand, Wm. McDougall, and others, he was a founder of the British Psychological Society in 1901. C. S. Myers became secretary in 1904 and Rivers served as one of the editors between 1904 and 1912. As early as 1910 a paper on "The Psychology of Freud and his School" was read before the Society by Bernard Hart. For further details see Beatrice Edgell, "The British Psychological Society," *British Journal of Psychology*, Vol. 37, 1947, pp. 113–132.

with Bernard Hart, Seligman, together with his wife, Brenda Z. Seligman, became vitally interested in the general relations of psychology to anthropology, in particular the bearing of psychoanalytic theories upon ethnological work. Fortes says that "he was dissatisfied with the concepts and methods of experimental psychology. Now he had an opportunity of experimenting with psycho-therapeutic methods based on new theories, especially those of psychoanalysis, which appealed strongly to his scientific sense. On taking up his anthropological work again after the war, Seligman turned these experiences to fertile use."[33]

The kind of work Myers and McDougall did on the expedition to the Torres Straits typified the psychological interests of the time. Writing a posthumous account of Rivers' work Myers later (1923) said: "Rivers interested himself especially in investigating the vision of the natives—their visual acuity, their color vision, their color nomenclature, and their susceptibility to certain visual geometric illusions. He continued to carry out psychological work of the same comparative ethnological character, after his return from the Torres Straits, in Scotland . . . during a visit to Egypt in the winter of 1900,

[33] Obituary notice in *Man*, Vol. 61 (1941), pp. 2, 4. Among the most important bibliographical items of the Seligmans that have a psychological interest are the following: C. G. Seligman, "Anthropology and Psychology: A Study of Some Points of Contact," *Journal of the Royal Anthropological Institute*, Vol. 54, 1924, pp. 13–46; "Anthropological Perspective and Psychological Theory," *ibid.*, Vol. 62, 1932, pp. 193–228; "Temperament, Conflict, and Psychosis in a Stone-Age Population," *British Journal of Medical Psychology*, Vol. 9, 1929, pp. 189–190. Brenda Z. Seligman, "The Incest Barrier," *British Journal of Psychology*, Vol. 22, 1932; "Incest and Descent," *Journal of the Royal Anthropological Institute*, Vol. 59, 1929; "The Part of the Unconscious in Social Heritage," in E. E. Evans-Prichard, ed., *Essays in Honor of C. G. Seligman*, London, 1934, pp. 307–317.

In the Introduction that Seligman wrote to J. S. Lincoln's *The Dream in Primitive Cultures*, London: The Cresset Press, 1935, he urged contemporary anthropologists to make themselves aware of psychological theory and asked: "How can a profitable give and take relationship between psychology and anthropology best be established at the present time? This is a problem," he goes on, "to which the writer of the Introduction has given much thought in recent years. Brought up in the main in the Tylorian (comparative) School of anthropology, having thereafter gained some knowledge of and made use of the Historical School of Rivers, and of late years watched the development of the functional method, the writer has become convinced that the most fruitful development—perhaps indeed the only process that can bring social anthropology to its rightful status as a branch of science and at the same time give it the full weight in human affairs to which it is entitled—is the increased elucidation in the field and integration into anthropology of psychological knowledge."

and from 1901–1902 in his expedition to the Todas of Southern India."[34]

What is of particular interest, now that we can view Rivers' work in the light of later developments, is the highly compartmentalized nature of his psychological interests, on the one hand, and his ethnological pursuits, on the other.[35] This lack of integration continued throughout his career, despite the fact that he no doubt aimed, as Lowie says, "to ally ethnology with psychology";[36] that he was influenced by Freudian theory in his later years,[37] and that, before becoming so thoroughly committed to the diffusionist theories of Elliot Smith and Perry, he raised some basic points in social psychology. When, for example, he asks in his 1916 essay on "Sociology and Psychology":[38] "How can you explain the workings of the human mind without a knowledge of the social setting which must have played so great a part in determining the sentiments and opinions of mankind?" he strikes a note that is far removed from his earlier specialized research in physiological psychology. This trend in his thinking is also exemplified when he dissents from Westermarck's view that it is more or less obvious that the blood feud can be explained everywhere by revenge by observing that so simple an answer leaves us about where we started. Rivers is of the opinion that

[34] Charles S. Myers, "The Influence of the late W. H. R. Rivers," in W. H. R. Rivers, *Psychology and Politics,* New York: Harcourt, 1923, pp. 158 ff.

[35] Elliot Smith recognized this. In the Preface to *Psychology and Ethnology,* New York: Harcourt, 1926, p. xiv, he says: "When Dr. Rivers embarked upon his first independent expedition [to the Todas] he worked both at psychology and at the investigation of religion and sociology. But the two lines of work were kept more or less distinct the one from the other. The psychological research was essentially physiological in nature and had no close or direct bearing on the other branch of his study the results of which were published in the volume *The Todas* (1906)."

[36] Lowie, 1937, p. 172.

[37] See *Instinct and the Unconscious,* 1922.

[38] In the *Sociological Review,* Vol. 9. Two years previously Rivers had expressed the opinion (*Kinship and Social Organization,* p. 3) that "the ultimate aim of all studies of mankind, whether historical or scientific, is to reach an explanation in terms of psychology, in terms of the ideas, beliefs, sentiments, and instinctive tendencies by which the conduct of man, both individual and collective, is determined." Rivers explains (1916) that by psychology he refers to the "science which deals with mental phenomena, conscious and unconscious." For psychologists, he says, he knows he begs the question by using the term "mental." What he wants to avoid, however, is "confusing mental processes with the social processes I regard as the subject matter of sociology." In particular, he is opposed to McDougall's broad definition of psychology in terms of behavior.

there are special questions for inquiry here: "Is revenge a universal human characteristic? Is it an emotion which has the same characteristics and the same content among all peoples, or does it vary with the physical and social environment? . . . An answer to one or more of these questions is suggested by some of the cases cited by Professor Westermarck, but he does not consider them from these points of view."[39] But such remarks of Rivers are often in the nature of *aperçus*. Although he touched upon vital issues,[40] he never initiated any serious research into questions of this kind.

Rejecting the approach of the classical evolutionists after his expedition to Melanesia, Rivers devoted more and more attention to the diffusionist theory sponsored by G. Elliot Smith and W. J. Perry. Personally, I have never been able to discover in his work any appreciation of the kinds of problems that later became the center of personality and culture studies, despite the knowledge he acquired of psychoanalytic theory. Nevertheless, the work in physiological psychology Rivers did in the Torres Straits, Egypt, and in India, provided substantial evidence for the first time that supported the hypothesis of psychic unity, considered at the psychophysical level. G. Murphy notes that Rivers' statement that "pure sense acuity is much the same in all races, has not been overthrown by subsequent research."[41]

Besides this, Rivers held the conviction that training "in the experimental methods of the psychological laboratory" was the "best training" for the field anthropologist, quite apart from special preparation for the investigation of psychological problems, as such. Rivers expressed this conviction to Sir F. C. Bartlett as a young man when the latter first went to Cambridge "somewhat undecided about a future career, but rather hoping to take up field work in anthropology." Many years later (1937) when Bartlett had become a distinguished psychologist he said that he was "sure that Rivers was

[39] "Sociology and Psychology," *op. cit.*
[40] Referring to the fact that "those explanations of custom which derive our economic scheme from human competitiveness, modern war from human combativeness, and all the rest of the ready explanations that we meet in every magazine and modern volume, have for the anthropologist a hollow ring." Ruth Benedict (*Patterns of Culture*, 1934, p. 214) points out that "Rivers was one of the first to phrase the issue vigorously. He pointed out that instead of trying to understand the blood feud from vengeance, it was necessary rather to understand vengeance from the institution of the blood feud."
[41] Gardner Murphy, *Historical Introduction to Modern Psychology*, revised edition, New York: Harcourt, Brace, 1950, pp. 359–360.

right" and goes on to express the personal opinion "that every anthropology student who hopes to make original contributions to his subject ought, at some period of his training—preferably in the later
stages—to spend six months or a year in the psychology laboratory,"
although he believes that his course should not be "identical with
that of the student who is to devote his life to research and teaching
in experimental psychology."[42]

While Bartlett followed a career in psychology rather than anthropology, he never lost interest in the latter subject.[43] Furthermore,
it seems reasonable to infer that the sociopsychological point of view
that he developed as a psychologist was related to his anthropological interests, and in particular to his association with Rivers. In his
Psychology and Primitive Culture (1923) he was asking:[44] "Are we,
in our search for explanations, always to go back to the individual
as he may be pictured to exist outside of any particular social group?
Shall we endeavour to find an origin for all forms of social behaviour
at some presocial stage? At first sight it may seem as if we are bound
to do this, if our study is legitimately to be called psychological. For
psychology is generally regarded as dealing essentially with the individual." Referring to Tylor and more particularly Frazer, he says
that the researches of the latter "yield constant attempts to account
for the absolute origin of rites and ceremonies in terms of individual
experiences, the individual not being considered specifically as a
member of any social group. . . ." This method, Bartlett goes on to
say, has been considered mistaken by other investigators. But "looking upon it as distinctly psychological, they propose to banish psychology from the study of primitive culture." Making references to
Kroeber's statement in his review of Freud's *Totem and Taboo* (1920)
that despite Frazer's "acumen his efforts are prevailingly a dilet-
tantish playing; but in the last analysis they are psychological, and
as history only a pleasing fabrication"; and that "Ethnology like every
other branch of science, is work, and not a game in which lucky
guesses score"; Bartlett remarks that, "there is no good reason for regarding either Sir James Frazer's work, or psychology, as a game of

[42] F. C. Bartlett, "Psychological Methods and Anthropological Problems,"
Africa, Vol. 10, 1937, pp. 401–420.

[43] His name appears in the *International Directory of Anthropologists*, 3rd
ed., Washington, D.C., 1950.

[44] Bartlett, 1923, pp. 8–11. Quoted by permission of the Cambridge University Press.

lucky guesses. But there is equally no good reason for regarding Frazer's type of explanation as the only type which can be called psychological. . . ." The psychologist is "not forced to consider only very remote antecedents, and with the curious exception of his studies in the realm of primitive culture, he never has agreed to do so. . . . In general, our problem is to account for a response made by an individual to a given set of circumstances *of which the group itself may always be one.* . . . The individual who is considered in psychological theory, in fact, is never an individual pure and simple. The statements made about him always have reference to a particular set of conditions. The individual with whom we deal may be the individual-in-the-laboratory, or the individual-in-his-everyday-working-environment, or—and in social psychology this is always the case—the individual-in-a-given-social-group."

This sociopsychological point of view links Bartlett and Rivers, although the latter never fully developed his ideas. Moreover, Bartlett says that the lectures on which his *Psychology and Primitive Culture* were based were undertaken "mainly on the advice" of Rivers; their subject matter was "many times discussed . . . with him" and, since Rivers had died before they were published, the author in paying tribute to Rivers says that he owes "more to him for his friendship and interest than I can adequately express."[45]

Radcliffe-Brown likewise studied psychology under Rivers. He says ". . . I was for three years his pupil in psychology, and was his first pupil in social anthropology in the year 1904. Rivers was from first to last primarily a psychologist, and was an inspiring teacher in psychology."[46] Radcliffe-Brown's profound interest in problems of social organization and his later contributions to this specialized field have obscured the fact that his first book, *The Andaman Islanders,* had a psychological focus. While not published until 1922 it was completed by 1914, and although the author expresses some dissatisfaction with Chapters 5 and 6 ("Interpretation of Andamanese Customs and Beliefs"), nevertheless it is these particular chapters he considers to be of special methodological importance. And they are the chapters in which the psychological focus is exemplified. "We

[45] *Ibid.,* Preface, p. viii.

[46] R. R. Radcliffe-Brown, "The Present Position of Anthropological Studies," Presidential Address, Sec. H, British Association for the Advancement of Science, 1930, p. 146. Reprinted in *Structure and Function in Primitive Society,* Glencoe: Free Press, 1953.

have to explain," he says,[47] "why it is that the Andamanese *think and act* in certain ways. The explanation of each single custom is provided by showing what is its relation to the other customs of the Andamanese *and to their general system of ideas and sentiments*" (italics ours). Chapter 5 deals with the psychological significance of ceremonies with special reference to their role in the collective expression and transmission of systems of "sentiments"; Chapter 6 with the part that myths and legends play "in the mental life of the Andaman Islander." Far from being "merely the products of a somewhat childish fancy" they "are the means by which the Andamanese express and systematize their fundamental notions of life and nature and the sentiments attaching to these notions."[48]

Radcliffe-Brown states five working hypotheses, all of which involve the concept of "sentiment." The first one is that "a society depends for its existence on the presence in the minds of its members of a certain system of sentiments, by which the conduct of the individual is regulated in conformity with the needs of society."[49] Consequently, in the case of the Andamanese, he is concerned with the demonstration that "the social function of the ceremonial customs of the Andaman Islanders is to maintain and to transmit from one generation to another the emotional dispositions on which the society (as it is constituted) depends for its existence."[50] It is made abundantly clear that the manner in which human beings are psychologically structured lies at the core of the persistence and functioning of a socio-cultural system. Besides this, although it may be postulated that there is "a general substratum that is the same in all human societies,"[51] the psychological organization of the individuals of one society as compared with another vary concomitantly with cultural differences: "a system of sentiments or motives will clearly be different in different cultures, just as the system of moral rules is different in societies of different types." Sentiments, moreover (as

[47] *The Andaman Islanders: A Study in Social Anthropology*, Cambridge University Press, 1922, p. 230.

[48] *Ibid.*, p. 330.

[49] *Ibid.*, pp. 233–234.

[50] In "Religion and Society," a lecture delivered in 1945 and printed in *Structure and Function*, Radcliffe-Brown reiterates this hypothesis. He says: "Thirty-seven years ago (1908) in a fellowship thesis on the Andaman Islanders (which did not appear in print until 1922) I formulated briefly a general theory of the social function of rites and ceremonies. It is the same theory that underlies the remarks I shall offer on this occasion."

[51] *Ibid.*, pp. 401 ff.

stated in hypothesis 3), "are not innate but are developed in the individual by the action of society upon him."[52]

Radcliffe-Brown's definition of sentiment—"an organized system of emotional tendencies centred about some object"—is that of A. F. Shand and William McDougall, although this is not explicitly stated.[53] Although the concept of sentiment is much too restricted a psychological construct for dealing with the complexities and varying patterns of affective and motivational phenomena that manifest themselves in the human personality, the broad psychological assumptions implied in Radcliffe-Brown's hypotheses are essentially the same as those which have appeared in personality and culture studies. In the latter it is likewise assumed that (a) however they may be construed or labelled, there are psychological constants that define the dynamics of a human level of adjustment everywhere; (b) any particular socio-cultural system is dependent for its persistence and functioning upon the characteristic psychological structuralization of its constituent members; (c) this psychological organization is acquired through social interaction and learning on the part of the individual; (d) it differs from one society to another; and (e) various institutions function with reference to the maintenance and expression of the affective and motivational structures of individuals. The major difference is one of research interest. Radcliffe-Brown is

[52] *Ibid.*, p. 234.

[53] *Ibid.*, footnote p. 234. McDougall integrated Shand's notion of sentiment with his own doctrine of instincts and contended that a theory of sentiments should be the foundation of a social psychology. (*Introduction to Social Psychology*, Boston: Luce, 1910.) He says (p. 122), "To such an organized system of emotional tendencies centred about some object, Mr. Shand proposes to apply the name 'sentiment'." E. Westermarck in his *Origin and Development of the Moral Ideas*, London: Macmillan, 1906, Vol. 1, p. 110, also used sentiment in the sense proposed by Shand. In his Huxley Memorial Lecture for 1932 ("Anthropological Perspective and Psychological Theory," *Journal of the Royal Anthropological Institute*, Vol. 62, 1932, pp. 193–228), C. G. Seligman said: "It is a remarkable fact that the concepts which have been laboriously achieved by the students of conscious behaviour seem to have been of little avail hitherto in anthropology. An examination of the writings of British anthropologists who attempt psychological explanation or analysis (Radcliffe-Brown, Malinowski, Brenda Z. Seligman) seems to show that the only concept derived from the psychology of conscious behaviour that has been of any service is that of the 'sentiment.' Is this to be attributed to the negligence of anthropologists, or rather to the fact that the psychological problems arising in anthropology lie for the most part not in the sphere of cognition—to which most attention has been paid in the psychology of consciousness—but in the sphere of motive and emotion, which is not intelligible without taking into account the findings of Freud, Jung, and a number of psychopathologists?"

interested in the function of concrete rites and ceremonies with reference to relatively abstract systems of sentiments. Culture and personality studies are concerned with discovering the actual psychological organization and dynamics of the human personality as it functions in different socio-cultural contexts, as well as the concrete factors in the life history of the individuals which are responsible for varying patterns of psychological structuring.

Although William McDougall, like Rivers, had been a member of the expedition to Torres Straits,[54] and later spent some time in Borneo, which led him to collaborate with Charles Hose on an anthropological book,[55] it is somewhat paradoxical that these field experiences led to no subsequent contacts of a fruitful nature with anthropology. Even though McDougall was well aware of the fact that nineteenth-century psychology had proved of little use to the social sciences, and pointed out in his *Introduction to Social Psychology* (1908)[56] that a "very important advance of psychology toward usefulness is due to the increasing recognition that the adult human mind is the product of the moulding influence exerted by the social environment, and of the fact that the strictly individual mind, with which alone the older introspective and descriptive psychology concerned itself, is an abstraction merely and has no real existence," nevertheless, his brand of social psychology provoked negative rather than positive reactions among anthropologists. Ogburn and Goldenweiser (1927)[57] remark that McDougall's instinct psychology did exert an early wave of psychological influence upon the social sciences generally, but that it was a "blind trail."

Before the end of the nineteenth century interest in what soon became known as "differential psychology" had begun to crystallize. This development represented a departure from the central problems of the experimental psychologists of this time as well as from the associationist tradition which was concerned with the general principles whereby "ideas" became associated, with no allowance for individual variations. Questions began to be raised about group differ-

---

[54] In his intellectual autobiography McDougall (*A History of Psychology in Autobiography*, C. Murchison, ed., 1930, p. 201) says: ". . . I was already interested in such topics as totemism, exogamy, and primitive religion, having read Tylor, Lang, Frazer, and other authorities in the field."

[55] *The Pagan Tribes of Borneo*, 2 vols., London: Macmillan, 1912. Mc-Dougall left for Borneo after having spent only five months in Torres Straits.

[56] P. 16.

[57] Ogburn and Goldenweiser, *The Social Sciences*, p. 4.

ences of various kinds, as well as differences among individuals. James McKeen Cattell (1860–1944) after his sojourn at Wundt's laboratory, had become interested in the measurement of individual differences. The term "mental test," referring to measures of ability at the sensory and motor level rather than intelligence, was first employed by him in 1890. What is of particular interest here is that Kroeber studied psychology under Cattell before he studied anthropology under Boas and that psychology was accepted as a minor for his doctorate.[58] Furthermore, two other anthropologists, Livingston Farrand and Clark Wissler, worked with Cattel in his early days at Columbia and published monographs[59] which Roback characterizes as a "harbinger of the avalanche of tests with which we were to be deluged a half-century later."[60] The subsequent application of intelligence tests in the exploration of racial differences in mentality has been critically reviewed by Klineberg, Nadel, and more recently by Anastasi and Foley.[61]

Meanwhile, William Stern (1871–1938) in Germany, had proposed the name "differential psychology" for a branch of the subject that would study "differences among individuals as well as among racial and cultural groups, occupational and social levels, and the two sexes" in a book first published in 1900.[62] Stern likewise founded an institute and a journal for applied psychology. In the latter there was published in 1912 a guide for the investigation of certain psychological problems among primitive peoples.[63] The anthropologist Richard Thurnwald wrote an introduction. The following year, the same journal published the results of Thurnwald's own field obser-

[58] A. L. Kroeber, *The Nature of Culture*, University of Chicago Press, 1952, p. 300.

[59] J. M. Cattell and L. Farrand, "Physical and Mental Measurements of the Students of Columbia University," *Psychology Review*, Vol. 3, 1896; Clark Wissler, "The Correlation of Mental and Physical Tests," *Psychological Monographs*, 1901, 3, No. 16, pp. 62 ff.

[60] A. A. Roback, *History of American Psychology*, New York: Library Publishers, 1952.

[61] Otto Klineberg, *Race Differences*, New York: Harper, 1935; Anne Anastasi and John P. Foley, Jr., *Differential Psychology: Individual and Group Differences in Behavior*, New York: Macmillan, 1949; S. F. Nadel, "The Application of Intelligence Tests in the Anthropological Field," in F. C. Bartlett, ed., *The Study of Society, Methods and Problems*, New York: Macmillan, 1939.

[62] Anastasi and Foley, *op. cit.*, p. 13.

[63] *Vorschläge zur psychologischen Untersuchung primitiver Menschen gesammelt und herausgegeben vom Institut für angewandte Psychologie.* Beihefte zur Zeitschrift für angewandte Psychologie und psychologische Sammelforschung, 5, Teil I, Leipzig, 1912.

vations made in 1906–1909, *Ethno-psychologische Studien an Süd-seevölkern auf dem Bismarck-Archipel und den Salomo-Inseln.*[64] In the earlier publication Thurnwald had maintained that the whole study of ethnology centers around the comprehension of the psychological peculiarities of non-European peoples. It is by this means he says that we really get to know them. Before he undertook his 1906 expedition Thurnwald had turned to Carl Stumpf (1848–1936) who consulted other psychologists. The result was an assembly of questions from a broad area of psychological interest which were to be tested for their utility in ethno-psychological research. This pioneer work on the part of Thurnwald did not, of course, include the use of standard intelligence tests, as later developed, or personality tests. But it did include a collection of drawings and verbal associations to selected vocabularies. Later (1925) Thurnwald founded the *Zeitschrift für Völkerpsychologie und Soziologie,* a journal which, in Lowie's words, "gave full scope to his interests, at once wide and deep, in economics, sociology, jurisprudence, and psychology" and made him "one of the foremost liaison officers of the social sciences."[65]

By the time anthropology had begun to assume the characteristics of an organized discipline in the United States, Wundt (1832–1921 was already a world-renowned figure. The effort he had made to establish psychology as an independent unified scientific discipline based upon experimental laboratory methods had borne fruit. Besides this, Wundt was unique among the psychologists of his period for his vision and unusual breadth of interests. His famous *Völkerpsychologie* was a product of the last twenty years of his life. Wundt, says Murphy,[66] "tries to bring together experimental psychology, child psychology, animal psychology, folk psychology; nothing that was psychology was foreign to him. He poured his energies into examination of nearly every corner of mental life." It

[64] *Ibid.,* 6, Leipzig, 1913. See also Thurnwald's later publication "Psychologie des primitiven Menschen" in Gustav Kafka, *Handbuch der vergleichenden Psychologie,* Vol. 1, Munchen, 1922. pp. 147–320.

[65] Lowie, *op. cit.,* p. 242. The following articles represent Thurnwald's point of view: "Probleme der Völkerpsychologie und Soziologie," *Zeitschrift für Völkerpsychologie und Soziologie,* Vol. 1, 1925, pp. 1–20; "Zum gegenwärtigen Stande der Völkerpsychologie," *Kölner Vierteljahrshefte für Soziologie,* IV, Jahrgang, neue Folge, 1925, pp. 32–43. See also F. H. C. Van Loon and R. Thurnwald, "Un questionnaire psycho-physico-morophologique pour l'étude de la psychologie des races," *Revue Anthropologique,* Vol. 40, 1930, pp. 262–277.

[66] Murphy, *op. cit.,* p. 159.

is of historical significance that he believed that cultural and historical data were relevant material for the psychologist. In consequence, he envisaged problems that centered upon the inter-relations
among the historic process, the individual, and varying cultural
forms and contents. Rejecting such terms as *Gemeinpsychologie*
and *Sozialpsychologie* he chose *Völkerpsychologie* as the term he
thought best suited to characterize this area of psychological interest
as Lazarus and Steinthal had already done. While it is quite true that
Wundt's folk psychology in terms of central focus and problems is
not equivalent to what later became known as social psychology
in America—it has been said that it represented an "heroic attempt
to save philosophy of history for science"[67]—nevertheless, the
kind of problems posed by Wundt, rather than his attempted solution
of them, are genuine. It is worth noting that in Kroeber's article
(1918) on "The Possibility of a Social Psychology"[68] he took as his
point of departure H. K. Haeberlin's article on Wundt.[69] Comment-
ing upon his own article in 1951 he says:

Of course, my "social psychology" of the time was something very different from the social psychology of today, which psychologists and sociologists have jointly reared as a discipline that deals with out-groups, mi-
norities, stereotypes, adjustments, attitudes, leadership, propaganda, public opinion, and other interpersonal relations within a society, without
reference to cultural content or forms except so far as a cultural content
may be needed to define a sociopsychological situation.
  What I was dreaming of, as a distant possibility, was a true processual
science causally explaining the pageant of the history of culture. I do not
see in 1951 that we are appreciably nearer such a science than we were in
1918. I still think that the processes through which cultural forms can be
explained causally must be largely or essentially psychological in nature.
In late years I have several times discussed "cultural psychology," though
as an inevitable psychological reflection or quality which all cultures must
bear rather than as their basic process or cause. I do not think that in and
by themselves the interpersonal or social relations of men in groups will
ever explain the specific features of specific cultures—certainly not their
ideologies, knowledge, skills, art, or understanding. Modern social psychologists do not even pose the problem. They are not interested in the

[67] Fay B. Karpf, *op. cit.*, p. 61, in the course of an extended discussion of
Wundt, pp. 51–65.
  [68] *American Journal of Sociology*, Vol. 23, pp. 633–651; reprinted in ab-
breviated form in *The Nature of Culture*, copyright 1952 by the University of
Chicago.
  [69] "Foundations of Wundt's Folk Psychology," *Psychology Review*, Vol. 23,
1916.

interrelation of cultural facts; just as I, in 1918, for all my talking about social psychology, was interested in culture and very little in social interrelations. In fact, next to culture, I was then and am now much more interested in the qualities and motivations of individuals than in what human beings do to one another in groups. The present paper must have puzzled both sociologists and psychologists at the time and since. I am doubly appreciative of their forbearance.[70]

According to Kroeber's conception of "cultural psychology" one of its tasks would be to proceed to the characterization of cultures empirically "without any set plan, merely noting those psychological traits which obtrude themselves in each culture, with special alertness toward such as seem to cohere into a consistent

[70] *The Nature of Culture,* p. 52. It should be borne in mind that in the 1918 article Kroeber stated that he conceived the "business" of the psychologist to be "the determination of the manifestations and processes of consciousness as consciousness." It was at this same period, too, that his "Eighteen Professions" (*American Anthropologist,* 17, 1915, pp. 283–288) and "The Superorganic" (*Ibid.,* 19, 1917, pp. 163–213), stirred up considerable dissent in regard to their psychological implications on the part of the three contemporary anthropologists who were most interested in psychology. It is not without significance, perhaps, that all of them were trained by Boas. Goldenweiser, in his chapter on "Cultural Anthropology" (in H. E. Barnes, *History and Prospects of the Social Sciences,* 1925) writes: "Kroeber's theoretical broadside was met by Sapir, Haeberlin, and me, who, while endorsing Kroeber's main contention as to cultural autonomy, took exception to his inadequate appreciation of the role of the individual in history, his over-confident assumption of historical determinism, as well as his theoretically inadmissible identification of psychology with biology." Haeberlin ("Anti-Professions," *American Anthropologist,* vol. 17, 1915, p. 759) had previously remarked that "as soon as Dr. Kroeber will have become conscious of the dogmatism of his biological psychology, all other obstacles towards an understanding must fall like a house of cards. He will recognize the impossibility of building a cloister-wall about history, he will no longer look askance on the psychologically inclined anthropologist as a hybrid form of two distinct crafts, psychology will no longer be a bugaboo—in short, there will be complete unison of the 'professions' and the 'anti-professions'." Nevertheless, Kroeber has continued to emphasize the sharp distinctions, as he conceives them, between the tasks of the psychologist and the anthropologist. In his *Anthropology* (New York: Harcourt, 1948) p. 572, while he points out that "modern psychologists recognize that the total picture of any natural or spontaneous human situation always contains a cultural ingredient," and that a " 'pure' unconditioned mind" is an abstraction, he says that "the business of psychologists, however, is to try to hold this cultural factor constant in a given situation, to account for or to equalize it, and then proceed to their own specific problems of investigating the mind *as if* it were 'pure,' of investigating the psychic aspects of the behavior of individual human beings, and after that man in general." Kroeber makes no reference, however, to any particular psychologists, or schools of psychological thought, nor to how psychologists themselves have conceived their central problem, either at present or in the past.

larger orientation."[71] Such an approach seems reminiscent in some respects of Wundt's idea, namely, a psychological history of man culturally viewed. At the end of his article[72] Haeberlin says: "What an intrinsic association of psychology and history can attain is well exemplified by numerous passages in Wundt's work on folk-psychology, when we abstract from all his theoretical foundations. There we find psychological interpretations of historical phenomena executed with a brilliancy characteristic of Wundt's genius. Such interpretations of Wundt's will mark the monumental significance of his work long after folk-psychology as such will have been recognized as [an] *'Unding'* "!

At the end of his chapter on "Cultural Psychology" Kroeber raises the question "whether the findings of a systematically developed cultural psychology will be expressed in the terms and concepts of individual psychology, or whether a new set of concepts will have to be added to these. . . ." This is precisely the problem that Wundt faced and it is the general consensus of opinion that despite his concepts of "appreciation" and "creative synthesis" he did not altogether escape the individualistic bias that prevailed.[73] On the other hand, Wundt did see the need for bringing the functioning of the "higher" mental processes in man into connection with the facts of culture even if psychologists were not prepared to handle these phenomena in the laboratory. He made an heroic effort to relate cultural data to psychological data at a time when the major emphasis in psychology was still individualistic, physiological, and biological. In the opinion of G. Stanley Hall his "doctrine of apperception . . . took psychology forever beyond the old associationism which had ceased to be fruitful"; he "established the independence of psychology from physiology" and "he materially advanced every branch of mental science and extended its influence over the whole domain of folklore, mores, language, and primitive religion."[74]

In any event, it would seem to be of some historical significance that it was Hall, a student of Wundt's, who was a key figure in the early promotion in the United States of anthropology and child psy-

[71] *Anthropology*, p. 621; cf. the previously unpublished paper "Culture, Events, and Individuals" (1946) in *The Nature of Culture*, pp. 104–109.

[72] Haeberlin, 1916, *op. cit.*

[73] Karpf, *op. cit.*, p. 53.

[74] Preface to American edition of Sigmund Freud, *A General Introduction to Psychoanalysis*, New York: Garden City Publishing Company, 1938, p. 6.

chology, as well as psychoanalysis.[75] He appears to have reflected his mentor's views regarding the importance of anthropology and ethnic psychology in relation to the study of the growth and development of the human mind when, after becoming President of Clark University in 1888, he invited Franz Boas there in connection with the program he envisaged for the development of genetic psychology. Boas stayed only a few years, but A. F. Chamberlain (1865–1914), the first student to receive a doctorate under Boas, remained. In addition to the anthropological field work he carried on among the Kutenai and the Mississauga Ojibwa, Chamberlain continued to function in the child psychology program at Clark. Roback, with no mention of his anthropological connection, refers to the fact that "in the early days of Clark, pedagogy and child psychology flourished under men like W. H. Burnham and A. F. Chamberlain."[76] His two books, *Child and Childhood in Folk-Thought* (1896) and *The Child: A Study in the Evolution of Man* (1900), received more attention from psychologists than anthropologists.[77]

In 1909 it was Hall who invited Freud, Jung, Jones, and other psychoanalysts, along with psychiatrists, psychologists, and biologists, to deliver lectures at the celebration of the twentieth anniversary of the opening of Clark University. On this occasion the subject of Boas' lecture was "Psychological Problems in Anthropology."[78] The purpose of it was "to point out a direction in which anthropological data may be used to good advantage by the psychologist." It is interesting to remark his concept of the scope of anthropology and the relation of anthropological data to psychological problems which Boas expresses in his opening paragraph.

The science of anthropology deals with the biological and mental manifestations of human life as they appear in different races and in different societies. The phenomena with which we are dealing are therefore, from one point of view, historical. We are endeavoring to elucidate the events which have led to the formation of human types, past and present,

[75] For Hall's "firsts" see Roback, *op. cit.*, pp. 154–155.

[76] Roback, *op. cit.*, p. 160.

[77] Chamberlain edited the *Journal of American Folklore* for almost a decade (1900–1908) and contributed for a considerable period a section to the *American Anthropologist* devoted to systematic abstracts of periodical literature.

[78] *American Journal of Psychology*, Vol. 21, 1910, pp. 371–384, embodied in *The Mind of Primitive Man*, 1st edition, New York: Macmillan, 1911. This volume of the *American Journal of Psychology* also contains lectures given by Freud, Jung, etc. on the same occasion.

and which have determined the course of cultural development of any given group of men. From another point of view the same phenomena are the objects of biological and psychological investigations. We are endeavoring to ascertain what are the laws of hereditary and of environmental variability of the human body. . . . We are also trying to determine the psychological laws which control the mind of man everywhere, and that may differ in various racial and social groups. Insofar as our inquiries relate to the last-named subject, their problems are problems of psychology, though based upon anthropological material.

So far as Boas' connection with Wundt is concerned, we know that, besides being familiar with his work, he brought Wundt to the attention of his students in his early years of teaching at Columbia. Writing thirty years later Benedict[79] reported that to Boas one of the central problems of anthropology "was the relation between the objective world and man's subjective world as it had taken form in different cultures."

Kroeber has pointed out that Boas "seems to have been wholly uninfluenced by Cattell, in spite of their long and close association on *Science* and at Columbia. He spent much time in his seminars, for several years, on Wundt; but it was the *Völkerpsychologie*, not the experimental psychology which Wundt helped found or organize, that occupied him."[80] Two early students of Boas took a special interest in Wundt. H. K. Haeberlin published the authoritative article already mentioned; A. A. Goldenweiser frequently referred to Wundt in his writings and, in addition, published an obituary notice in *The Freeman* (1921) upon the occasion of Wundt's death.[81]

Goldenweiser maintained (1922) that Wundt represented a psychological position greatly in advance of that reflected in the work of such English evolutionists as Spencer, Tylor, or Frazer, who were still deeply immersed in the individualistic, rational, and empirical tradition of the associationistic school. Wundt, says Goldenweiser, "discarded the crude rationalism of Spencer and Tylor. To him early man was not an aboriginal thinker facing nature as a set of problems or questions to which animism or magic could provide an answer or

[79] "Franz Boas, 1858–1942," A. L. Kroeber, Ruth Benedict, Murray B. Emereau, et al., *American Anthropologist*, Vol. 45, no. 3, Part II, 1943.

[80] *Ibid.*

[81] The major references to W. Wundt's theories are *Early Civilization*, 1922, reprinted in *History, Psychology and Culture*, 1933; "Psychological Postulates of Wundt's Folk-Psychology," a section of "Anthropology and Psychology," in the *Social Sciences and Their Interrelations*, edited by Ogburn and Goldenweiser, 1927, also reprinted in *History, Psychology and Culture*, pp. 77–80.

solution. Wundt saw clearly that man's reactions to the world—and especially his earliest reactions—were least of all rational and deliberate; rather were they spontaneous and emotional. The associationism of Frazer also collapsed before Wundt's doctrine of apperception, in which the atomistic and analytical view of mind was supplemented by an approach in which its integrative and creative functions were emphasized. Again, Wundt realized that the psychological foundations of civilization cannot be sought in the isolated individual, but that the group always actively cooperated in the production of attitudes and ideas. With great erudition and an originality that has often been underestimated, Wundt examined from this general standpoint the phenomena of language, art, religion, and mythology, social organization and law. Without espousing the doctrine of a separate folk-soul—a doctrine sponsored, e.g., by such German philologist-philosophers as Steinthal and Lazarus—Wundt insisted that civilization was impossible without the interrelated experiences of individuals in communication. Without laying himself open to the accusation of over-emphasizing the social—the principal weakness of the Durkheim school—Wundt joined the ranks of most modern sociologists and ethnologists in stressing the social and cultural setting."[82] Some of these same points were, of course, those made by Lévy-Bruhl who, in his earliest and most famous book (1910) directly challenged the psychological assumptions of the English evolutionists.[83] And Malinowski, despite his generally laudatory attitude towards Frazer, is fully aware of the deficiences of the kind of psychological interpretations for which Frazer was famous.[84]

In 1927, and even later, Goldenweiser expressed the view that

[82] *History, Psychology, and Culture,* copyright by Alfred A. Knopf, Inc., pp. 189–190.

[83] *Les fonctions mentales dans les sociétés inférieures;* English translation, *How Natives Think,* London: Allen and Unwin, 1926, Introduction.

[84] Malinowski, *A Scientific Theory of Culture and Other Essays,* Chapel Hill: University of North Carolina Press, 1944, p. 188.
"Frazer was essentially addicted to psychological interpretations of human belief and practice. His theory of magic, as the result of association of ideas; his three consecutive hypotheses about the origins of totemism in terms of belief in 'external soul,' 'magical inducement of fertility,' and in 'animal incarnation,' are essentially conceived in terms of individual psychology. Those who know his treatment of taboo, of the various aspects of totemism, of the development of magic, religion, and science, well realize that, throughout, Frazer in his explicit theories is little aware of the problems of social psychology. He is as already mentioned, fundamentally hostile to psychoanalysis, while behaviorism never enters his universe of discourse."

Wundt still had psychological meat for the contemporary anthropologist and other social scientists.[85] In particular, he calls attention
to the value of Wundt's conception of the "mutation of motives" and
the "heterogeny of ends," as applied in the study of the dynamics of
culture.[86] Nadel (1951) finds the latter idea useful, if (as he says) we
disregard the special evolutionary implications which Wundt gave it.
For Nadel, "it corrects the slight presumption of adequacy which
goes with the word 'integration' and it also emphasizes the fluidity
and the dynamic character of this integration of purposes, which is
constantly 'becoming' and never final."[87]

From 1900 onward, says Flugel, "we find psychology embarking
on the process of specialization incidental to the growth of new
schools, each school having its own peculiar methods and outlook
and even to a considerable extent its own peculiar jargon, so that
the total picture becomes one of increasingly violent and bewildering
activity; eventually, indeed, there are 'psychologies' rather than
'psychology' and students begin to complain that what is taught in
one center bears little resemblance to that which they have learned
in another. It is the period of psychoanalysis, behaviorism, and
Gestalt, of mental tests and the psychology of individual differences,
of 'factors,' of reflexology, and of the application of psychological
methods and concepts to the hither-to quite foreign fields of in-
dustry and commerce."[88]

Setting aside the influence derived from psychoanalytic psychology and psychiatry for the moment, we may ask: What were the influences of behavioristic psychology and Gestalt psychology on
anthropology? A closer scrutiny of the facts than it is possible to
give here would, I believe, show that the influence of behaviorism

---

[85] And Malinowski, *Scientific Theory*, p. 25, refers to the volumes of Wundt's
Völkerpsychologie as among the anthropological works which "command our
respect and admiration."

[86] Goldenweiser, *History, Psychology, and Culture*, copyright by Alfred A.
Knopf, Inc., pp. 79–80.

"This contribution of Wundt's to the dynamics of cultural life went over
the heads of the evolutionists, the diffusionists shut their doors . . . against it,
and even the critical anthropologists who should have known better, were too
busy disposing of their predecessors to do it justice. It is to be hoped that the
superior discernment embodied in Wundt's concept will not be wasted on social
thinkers during the present period of mutation of motives and purposes in the
entire field of the sciences of society."

[87] S. F. Nadel, *The Foundations of Social Anthropology*, Glencoe: Free Press,
1951, p. 386.

[88] Flugel, *op. cit.*

far surpasses that of any other school of psychological thought. Nor is it difficult to understand why. As pointed out by Bruner and Allport, the period 1908–1918 was the one characterized by behaviorism and objectivistic methods and aims. This is also the period when anthropology in America was getting well under way. By this time anthropologists knew so much about cultural variability and were so fascinated by the questions raised that any psychology that was primarily concerned with generic innate traits of man appeared to have no reference to the vital problems of the anthropologist. Thus Lowie, writing in 1917, asserted that "culture is, indeed, the sole and exclusive subject matter of ethnology, as consciousness is the subject matter of psychology," and went on to say that "the science of psychology, even in its most modern ramifications of abnormal psychology and the study of individual variations, does not grapple with *acquired* traits nor with the influence of *society* on individual thought, feeling, and inclination. It deals on principle exclusively with *innate* traits of the individual."[89] In 1940 Klineberg, referring to this statement of Lowie's, says that it has "now a somewhat antiquated flavor."[90] However, if the characterization of psychology expressed by Boring is recalled, and consideration is given to the fact that mental tests were being focused upon the discovery of "native" intelligence and that, in early child psychology, "natural" and "universal" stages in ontogenetic development were stressed, often colored by the recapitulation theory under Hall's influence, and it is not forgotten that it was the behaviorists who greatly stimulated the study of learning and that personality and social psychology had, as

[89] R. H. Lowie, *Culture and Ethnology*, New York: Douglas C. McMurtrie, 1917, pp. 5, 16. Clark Wissler in his address to the AAAS, as Chairman of Section H, 1915 ("Psychological and Historical Interpretations for Culture," *Science*, Vol. 43, 1916, pp. 193–201) expressed a similar viewpoint. "Psychologists give their attention to innate phenomena," he said, "especially man's psycho-physical equipment. If we extend the meaning of the term behavior so as to include consciousness, we may say that psychologists are concerned with the behavior of man as an individual." Wissler goes on to say that while "it may be that there is a problem in the comparative behavior of the individuals comprising ethnic groups . . . it is a psychological one and must be solved by the use of psychological data. Anthropologists give it little concern because they see in differences of individual behavior no significant cultural correlates." Known cultural phenomena since the paleolithic "necessitate no change in man's innate equipment nor in his innate behavior. So, on the whole, anthropology is quite indifferent to the problem of comparative behavior because it is concerned with the objective aspects of what is learned in life."

[90] Otto Klineberg, *Social Psychology*, New York: Holt, 1940, p. 5.

yet, no scientific respectability, Lowie's impression of the central interest of psychology in America in the second decade of this century can be easily appreciated.

Behaviorism had an appeal for the anthropologists because it, too, was opposed to the prevailing doctrine of instincts (substituting for them more specific physiological needs or drives), innate mental differences, or innate anything. Watson practically guaranteed to make a specialist in any of the professions or arts out of a healthy infant "regardless of his talents, peculiarities, tendencies, abilities, . . . or race of his ancestors."[91] This extreme environmentalist doctrine fitted in very well with the idea that culture was acquired, and that individuals with different cultural backgrounds acquired different sets of habits. The behaviorists promoted the study of learning but, says Asch,[92] "adopted a particular account of the learning process founded in the old doctrine of association and elaborated in the investigations of conditioned responses. Dominant in this concept were the role of *trial* and *error*, and the operations of *reward* and *punishment*. This formulation of the learning process was based primarily on the interpretation of the problem-solving activities of infra-human organisms but was given a general application and extended to the human level." Soon the term "conditioning," one of the verbal earmarks of the behaviorist, became almost as familiar in anthropological as in psychological literature, even though the purely technical psychological meaning of it and the theory behind it remained unexamined. Wissler remarked in 1923 that ". . . anthropologists are, in the broad sense, behaviorists, and so stand shoulder to shoulder with those psychologists to whom the same term applies. . . ."[93] And

[91] J. B. Watson, *Behaviorism,* New York: Norton, 1925, p. 82. Referring to the psychological scene in the years following 1910 Boring (*History of Experimental Psychology,* p. 494, first edition) says:

"Some conservatives were Wundtians, some radicals were functionalists, more psychologists were agnostics. Then Watson touched a match to the mass, there was an explosion, and behaviorism was left. Watson founded behaviorism because everything was all ready for the founding. Otherwise, he could not have done it. He was philosophically inept, and behaviorism came into existence without a constitution. Ever since, the behaviorists have been trying to formulate a satisfactory epistemological constitution and thus to explain themselves."

[92] Solomon E. Asch, *Social Psychology,* copyright 1953 by Prentice-Hall, Inc., New York, p. 12.

[93] Clark Wissler, *Man and Culture,* New York: Crowell, 1923, p. 251. At the same time the hypothesis Wissler developed to account for a "universal pattern" of culture is a theory involving psychobiological determinism. He postulates an innate cultural drive since he says (pp. 264–265) "that the pattern for

Nadel (1951) says: "A few sociologists and anthropologists have fully accepted the tenets of behaviourism; many more make concessions to it. If the use of the term 'conditioning' is any evidence, almost the whole of modern anthropology has gone behaviourist. But mostly this and similar verbal concessions are only just that—whether they result from lip service or eclecticism. In either case the essence of behaviourism has been misunderstood: it offers a challenge, not an expedient." "Certainly, behaviouristic psychology in its early and extreme form was as naive as it was ambitious," says Nadel, ". . . moreover, most behaviourists would claim that they have outgrown the crudeness and assurance of this early approach." But according to him, "some anthropologists . . . still subscribe to it. The textbook on anthropology by Chapple and Coon (1942) is an example. G. P. Murdock's article on 'The Common Denominators of Culture' at least savours of 'primitive' behaviourism."[94]

In Nadel's opinion "the earlier claims of behaviorists to be all-embracing are only slightly toned down" even in the later system of Hull to which specific reference is made. Hull was not only an important figure as one of the few theoreticians that America has produced, but because at Yale he directly influenced such anthropologists as Murdock, Gillin, Ford, and Whiting. As compared with other anthropologists who may have used the vocabulary of behaviorism or been influenced by general behavioristic concepts, this group is fully aware of the postulates and implications of Hull's system in all their ramifications.

Although it would be difficult to prove it, part of the early resistance to psychoanalytical psychology and the study of personality in culture, was probably due to the entrenchment by that time of behaviourism as the sort of psychology that was objective and scientific, as compared with psychoanalysis which again raised the ghost of the innate dynamic tendencies in man in a new form and seemed to have a subjective aura about it.

Malinowski's final and systematic exposition of his views is an

---

culture is just as deeply buried in the germ plasm of man as the bee pattern in the bee" so that "a human being comes into the world with a set, or bias, to socialization, according to a definite pattern, . . . by reason of which a man is a human being and not a termite, a bee, nor even a monkey. The human pattern, therefore, is a part, if not the whole, of man's inborn behavior . . . man builds cultures because he cannot help it, there is a *drive* in his protoplasm that carries him forward even against his will."

[94] Nadel, *op. cit.*, pp. 57, 59.

outstanding example of a considered approach to culture in behavioristic terms. It is more important to recognize this fact than to fall back upon his own self-labelling and call him a "functionalist." For Malinowski is quite explicit about his psychological position. He says that "the approval of psychoanalysis does not in any way detract from the great importance which behaviorism promises to acquire *as the basic psychology for the study of social and cultural processes* [italics ours]. By behaviorism I mean the newer developments of stimulus and response psychology as elaborated by Professor C. Hull at Yale, Thorndike at Columbia, or H. S. Liddell at Cornell. The value of behaviorism is due, first and foremost, to the fact that its methods are identical as regards limitations and advantages with those of anthropological field work. In dealing with people of a different culture, it is always dangerous to use the short-circuiting of 'empathy,' which usually amounts to guessing as to what the other person might have thought or felt. The fundamental principle of the field worker, as well as of the behaviorist, is that ideas, emotions, and convictions never continue to lead a cryptic, hidden existence within the unexplorable depths of the mind, conscious or unconscious. All sound, that is experimental, psychology can deal only with observations of overt behavior, although it may be useful to relate such observations to the shorthand of introspective interpretation."[95]

The implications of this behavioristic approach, shared by anthropologists other than Malinowski, have been sharply generalized by Asch.[96] "Culture becomes in this view a superstructure of habits and tendencies, 'a vast conditioning apparatus' geared to the gratification of primary needs. The prime movers of action are the biological needs; these provide the energy for all other psychological processes. All modifications of behavior, in particular social motives and practices, take place in relation to the gratification of primary needs. . . . We find here also the bold claim that the fact of society, the course of history, the growth of thought have introduced no new ends, that for all man's achievements his ends remain the same as those of infra-human organisms. He differs from these solely in the possession of a superstructure of an elaborate set of tools, material and psychological. Human activities are either consummations of primary needs or means to their consummation. If we find in society

[95] Malinowski, *op. cit.*, Chapel Hill: University of North Carolina Press, 1944, p. 23; for Nadel's comments on Malinowski's position see *Foundations*, p. 378.

[96] Asch, *op. cit.*, copyright 1953 by Prentice-Hall, Inc., New York, p. 14.

what appear at first sight to be other needs or ends, say for companionship, or knowledge, we must trace them to their instrumental value for primary needs. There can be nothing new in the way of purpose and aspiration. The crux of the doctrine is to deny this possibility; there can only be more circuitous ways of satisfying primary needs. As one noted anthropologist [Malinowski] has stated: 'By human nature, therefore, we mean the biological determinism which imposes on every civilization and on all individuals in it the carrying out of such bodily functions as breathing, sleep, rest, nutrition, excretion, and reproduction'."[97]

It may be noted in passing that the reductionistic implications of such a behavioristic approach to culture have their parallel in Freud's treatment of culture in *Civilization and Its Discontents*. Furthermore, it is just this reductionism, and its implications for personality theory as well, that represents one of the central points of difference between Jung and Freud and which led to the break between them. In a broad and fundamental sense, Freud, as well as many psychiatrists, are within the behaviorist tradition of psychological thinking.

The use of such concepts as "configuration" and "pattern," as applied to a culture, do not actually indicate any direct or profound influence of Gestalt psychology upon anthropology. They are even more superficial indications of genuine influence than is the frequent occurrence of the term "conditioning" in anthropological literature a clue to what a systematic behavioristic approach means. Kroeber, for instance, introduces his chapter on "Patterns" by observing that they "are those arrangements or systems of internal relationship which give to any culture its coherence or plan, and keep it from being a mere accumulation of random bits. They are, therefore, of primary importance. However, the concepts embraced under the term 'pattern' are still a bit fluid; the ideas involved have not yet crystallized into sharp meanings. It will therefore be necessary to consider in order several kinds of patterns."[98] No references are made to *Gestalt* psychology in this chapter, but in a section of the previous one, entitled "Content and Form; Ethos and Eidos; Values," attention is called to the elementary point that "a system or configuration is always, in its nature, more than the mere sum of its parts; there is also the relation of the parts, their total interconnections,

[97] The quotation is from Malinowski, *op. cit.*, p. 75.
[98] Kroeber, *Anthropology*, New York: Harcourt, 1949, p. 311.

which add up to something additionally significant. This is well recognized in 'Gestalt' or configurational psychology. The 'form' of culture may therefore be regarded as the pattern of interrelations of the contents that constitute it."[99]

Honigmann, in his summarization of the literature on "the configuration approach," makes no reference to Gestalt psychology, but stresses the fact that such pioneers as Sapir and Benedict "followed the leads of German historical philosophers like Dilthey and Spengler, who had already approached conceptualization of similar phenomena in their *Weltanschauung* and *Zeitgeist* studies."[100] Sapir had employed the term "spirit" or "genius" in his famous article "Culture, Genuine or Spurious" (1924).[101] In her earliest articles on "Psychological Types" (1930)[102] and "Configurations of Culture in North America" (1932)[103] Benedict made no reference to Gestalt psychology. In her *Patterns of Culture* (1934) she gives explicit credit to Dilthey and Spengler for recognizing "the importance of integration and configuration" *in cultures,*[104] giving only passing mention to Gestalt psychology.[105] Boas, in his introduction to her book, calls attention to the fact that the author "calls the genius of culture its configuration." Long before this, however, Boas himself had been interested in the same question. For in writing the obituary of H. K. Haeberlin (1919)[106] in which he emphasizes the "keen psychological interest of the latter" Boas observes that ". . . the wider concept of culture as dominating all the phases of tribal life occupied his attention. . . . With remarkable clearness of vision . . . he grasped the psy-

---

[99] *Ibid.,* p. 293.

[100] John J. Honigmann, "Culture and Ethos of Kaska Society," New Haven: *Yale University Publications in Anthropology,* No. 40, 1949, p. 10. Goldenweiser discussing the influence of Wilhelm Dilthey expresses the opinion that "of modern movements, *Gestalt* psychology and configurationism in anthropology without articulating directly with Dilthey are obviously related to him in intellectual orientation. . . ." ("The Relation of the Natural Sciences to the Social Sciences" in *Contemporary Social Theory,* edited by H. E. Barnes, Howard and Frances B. Becker, New York: Appleton-Century, 1940, p. 93, footnote.)

[101] *American Journal of Sociology,* Vol. 29, 1924, pp. 401–429.

[102] *Proceedings of the 23rd International Congress of Americanists,* New York: 1930, pp. 572–581.

[103] *American Anthropologist,* Vol. 34, 1932, pp. 1–27.

[104] *Patterns of Culture,* Penguin edition, p. 47.

[105] She observes (*ibid.*) that while psychologists of this school have worked "chiefly in those fields where evidence can be experimentally arrived at in the laboratory . . . its implications reach far beyond the simple demonstrations which are associated with its work."

[106] *American Anthropologist,* Vol. 21, 1919, pp. 71–74.

chological basis of culture as a unit." From Haeberlin's "Idea of Fertilization in the Culture of the Pueblo Indians.[107] Boas quotes the statement "that culture is not comprehensible as a summation of diffused elements is proved by the re-interpretation of heterogeneous traits according to a uniform scheme of interrelated ideas. The problem of the cultural setting of the Pueblos is therefore a psychological one."

What Benedict really brought into focus initially, therefore, was the question of how we are to conceptualize and study the "wholeness" of cultures, in contrast with an analytic approach that made primary use of such concepts as culture "traits" and "complexes."[108] She is dissatisfied with functionalism. She says, "Malinowski, somewhat disappointingly, does not go on to the examination of these cultural wholes, but is content to conclude his assignment with pointing out in each context that each trait functions in the total cultural complex, a conclusion which seems increasingly the beginning of inquiry rather than its peroration. For it is a position that leads directly to the necessity of investigating in what sort of a whole these traits are functioning and what reference they bear to the total culture. In how far do the traits achieve an organic interrelation? Are the *Leitmotivs* in the world by which they may be integrated many or few? These questions the functionalists do not ask." It is in this article, too, that Benedict says that her prime characterizations of psychological types in the Southwest have referred to the ethos of these people and that "cultural configurations stand to the understanding of group behavior in the relation that personality types stand to the understanding of individual behavior."[109]

The question of culture-wholes is thus related to the concept of "ethos" and this connection is brought out both by Honigman and

[107] *American Anthropological Association, Memoir,* 3, 1916.

[108] Cf. John Gillin, "The Configuration Problem in Culture," *American Sociological Review,* Vol. 1, 1936, pp. 373–386, and Clyde Kluckhohn, "Patterning as Exemplified in Navaho Culture" in *Language, Culture, and Personality: Essays in Memory of Edward Sapir,* L. Spier, et. al., eds., Menasha, Wisconsin: Banta, 1941. More recently John Gillin and George Nicholson ("The Security Functions of Cultural Systems," *Social Forces,* Vol. 30, 1951, pp. 179–84) have offered a mathematical model for dealing with the security functions of cultures considered as wholes. The concept of ethos is introduced with reference to an index that "would be a measure of the quality of the way in which a culture defines threats."

[109] Benedict, *Configurations,* pp. 2, 23. There is a reference in this article to William Stern's book on the human personality (1919).

Kroeber. The further step which Benedict took was to translate the "patterns" or "configurations" of the cultures of peoples into psychological terms. Whiting and Child epitomize the problem upon which Benedict focused attention in *Patterns of Culture* by saying that she "shows that many of the major aspects of a culture fit into a pattern or configuration which may be described in terms of motivational orientation or personality type. This basic orientation is held to have an important selective effect in the adoption, development, and modification of all sorts of specific aspects of the culture. The basic orientation, of course, Benedict believes to be culturally rather than biologically transmitted; in this book she is not concerned with the effect of culture on personality but with the selective effect on culture which this orientation has, once it has been established."[110] While this approach did not involve the use of Gestalt psychology in any technical sense, it represents a noteworthy parallelism, in principle, between the rejection of an older "elementaristic" approach on the part of the Gestalt psychologists and a similar movement away from an analytic approach to culture on the part of some anthropologists. Asch has called attention to this analogy,[111] and concludes that "as a result elementarism in anthropology gave way to the more fruitful assumption that the unit of investigation is the society." Nadel, who has acknowledged a debt to Gestalt psychology,[112] discusses a whole series of problems inherent in any approach to cultures or societies as wholes ranging through ethos and eidos to culture and personality. In his opinion "the theory of culture patterns has a mixed ancestry. It subscribes to the tenet of Gestalt psychology that the 'whole is more than its parts'; it draws on such popular conceptions as the 'spirit of the times' or the 'genius' of a people or civilization; it also harks back to Nietzsche's early philosophy and the German 'verstehend Psychologie'; and it claims some kinship with the theories of personality."[113]

Even if it be acknowledged that there is a significant analogy between the anthropologists' interest in cultures conceived as integral functioning wholes and the Gestalt idea, we may ask: What is the direct relevance of Gestalt psychology to anthropological re-

---

[110] John W. M. Whiting and Irvin L. Child, *Child Training and Personality: A Cross-Cultural Study*, New Haven: Yale University Press, 1953, p. 2.

[111] Asch, *op. cit.*, copyright 1953 by Prentice-Hall, Inc., New York, p. 61.

[112] Nadel. *op. cit.*, Preface, p. vi.

[113] *Ibid.*

search, and what does the record show? To what extent have anthropologists actually made use of the principles, methods, and concepts of Gestalt psychology in their work? What is the relevance, for instance, of the principles of this school to those areas of anthropological research where, instead of the primary concern being with patterns of culture or ethos, attention is focused upon the differential behavior of human beings as members of different socio-cultural systems? The manner in which the individual responds to the meaningful aspects of the world in which he has to live and act, his perceptions and his motivations, in short, the "psychological field" constituted for him by his membership in a group, certainly suggests the relevance of the kind of approach that Gestalt psychologists have developed. Nevertheless, compared to the influence exerted by behaviorism, Gestalt psychology seems to have had only the most superficial effects upon anthropological research. This is particularly noteworthy since two of the founders of the movement have themselves manifested an interest in the use of ethnographic data. Wertheimer's discussion of number concepts in the thinking of primitive peoples, although originally published in 1912,[114] does not appear to have interested anthropologists, at least in the United States. And while Köhler's *Mentality of Apes* is universally known, his "Psychological Remarks on Some Questions of Anthropology" (1937)[115] has gone practically unnoticed. It is in this article that Köhler sagely observes that "though some anthropologists do not like to admit it, psychological principles play an important role in the interpretation of anthropological facts. Theoretical difficulties may, therefore, arise quite as easily from inadequate psychological notions as from the strange ways of primitive mentality." And toward the close of the article he observes that although "hardly a word has been said about primitive religion, none about primitive art, again none about social life with all its institutions and ramifications, I am convinced that in these fields, too, psychology can be of more help now than, say, thirty or even twenty years ago. But for such help other principles besides those of the present essay will have to be introduced."

There are several reasons why anthropologists have not been immediately attracted to Gestalt psychology. In the first place, the rele-

[114] Max Wertheimer, "Über das Denken der Naturvölker: Zahlen und Zahlgehelde," *Zeitschr. f. Psychologie*, 60, 1912. Translation in Willis D. Ellis, *A Source Book of Gestalt Psychology*, New York: Harcourt, 1938.

[115] *American Journal of Psychology*, Vol. 50, 1937, pp. 271–288.

vance of this approach can only become significant for the anthropologist who is interested in behavior and not simply in culture *per se* or even cultural configurations abstracted from human behavior. Besides this, both Wertheimer and Köhler are particularly interested in cognitive processes at the level of psychodynamics, and not in culture and personality. Furthermore, since the general point of view of Gestalt psychology is antithetical to the radical environmentalism that characterized behaviorism in its early stage, any dogmatic form of cultural determinism is incongruous with the Gestalt approach. In the article referred to, Köhler, for example, explicitly rejects the assumption that "the individual is no more than an empty container for the products of group mentality," the same point which was raised by Sapir, Goldenweiser, and Haeberlin in their criticism of Kroeber and which is even more pertinent with reference to the "culturology" of Leslie White.[116] Finally, while the approach of Gestalt psychology is particularly relevant to the study of the complex relations existing between the organization of the psychological and social fields of the individuals of one society as compared with another, which may also, in part, be considered functions of differential cultural factors, a relativistic viewpoint is not maintained.[117] For instance, it is Köhler's position that while magical beliefs may differ from one society to another, and while the individual derives his beliefs from his group membership, nevertheless, "to some extent magic exists in practically every society . . . [and] at least some of its major premises are not peculiar to a few specific tribes but are the common property of all mankind below a certain high level of sophistication. From this common stock, which general developmental psychology is entitled to study and to explain, different

116 As, for example, when White writes ("The Individual and the Cultural Process" in *American Association for the Advancement of Science, Centennial Volume*, 1950, pp. 74–81) that the individual is "merely an organization of cultural forces and elements that have impinged upon him from the outside and which find their overt expression through him. So conceived, the individual is but the expression of a suprabiological cultural tradition in somatic form," or that "relative to the culture process the individual is neither creator nor determinant; he is merely a catalyst and a vehicle of expression." Perhaps the most extreme epitomization of this position is White's statement ("The Locus of Mathematical Reality: An Anthropological Footnote," *Philosophy of Science*, Vol. 14, 1947, p. 296) that after all, possibly "the most effective way to study culture scientifically is to proceed *as if* the human race did not exist."

117 See, e.g., Chapter 13 in Asch, *op. cit.*, "The Fact of Culture and the Problem of Relativism."

societies, with their different environments and histories, have in fact devised different varieties of actual practices. I do not believe that we can fully understand the origin of such varieties before we know on what ground magic in general grows." With the increasing interest that is developing in constancies and cultural universals, the point of view of Gestalt psychology in regard to such problems may prove more relevant to anthropology in the future than it has in the past.[118]

Personally, I have found Koffka's concept of a "behavioral" as differentiated from a "geographic" or objective environment of particular usefulness in defining the organism-environment relations at the human level.[119] From a psychological standpoint the environment of man is always a culturally constituted behavioral environment. Gestalt psychologists have also stressed self-perception as an essential constituent of the structuralization of the psychological and social fields of the human being during a period when behaviorists and other academic psychologists avoided any discussion of the self. In "The Self and Its Behavioral Environment" I have discussed the relevance of cultural variability in the self-image and self-con-

[118] Donald N. Michael ("A Cross-Cultural Investigation of Closure," *Journal of Abnormal and Social Psychology*, 48 (1953), pp. 225–230), for example, wishing to determine whether closure may be considered "a general law of innate perceptual organization" or whether "differences in cultural conditioning" influence the perception of closure, set up an experiment in which he tested the responses of Navajo and white subjects with respect to the detection of "small openings in tachistoscopically presented circular stimuli." Despite the well known fact that the Navajo are motivated to non-closure in the execution of certain ornamental designs by the values of their culture, no significant differences in the perception of closure were found in the two groups of experimental subjects.

[119] For a more extended discussion, see "The Self and Its Behavioral Environment" in *Explorations*, Feb. 1954 (Toronto, Canada). Without systematic qualification, or further analysis, the term "environment" has long proved ambiguous and unsatisfactory. The distinction made by Koffka is followed by other Gestalt psychologists, although they may use a different terminology. Lewin contrasts "objective environment" or "foreign hull" of the life space and "psychological" environment; David Krech and Richard S. Crutchfield (*Theory and Problems of Social Psychology*, New York: McGraw-Hill Book Company, Inc., copyright 1948, p. 38) write: "The real environment of a person is that environment which would be described by an objective observer; the psychological environment is that which would be described by the experiencing person himself. . . . The very same physical environment can result in radically different psychological environments for two different persons." The same statement may be generalized for what I have called "the culturally constituted behavioral environment" of different *groups* of mankind.

cepts of different peoples in relation to basic orientations that appear to be primary functions of all cultures, once it is recognized that self-awareness is a product of the socialization process and a primary constituent of a human personality structure essential for the functioning of any human society.

Asch, approaching social psychology systematically from a Gestalt point of view, makes a further point. He insists that other psychological schools of thought, no matter whether they have taken the form of environmental or instinct doctrines, have failed to clarify "the invariant or the mutable characteristics of men. . . . Traditional instinct and habit psychologies have little of relevance to say about the specific properties of human orientation; they give the appearance of having solved the problems of social life when they have simply bypassed them. In particular, they do not face seriously the problems of order in individual and social action when they describe men as a sum of instincts and habits. A psychology of drives and habits can hardly find a conceptual place for the psychological structures most characteristic of man—for the reality of a self, of kinship relations, or a sense of values. It results in the description of an individual who is not capable of novelty, who is not a genuinely social being but is only a more complicated form of a pre-social, pre-human individual. It is a crude fallacy to assume that no changes occur in the social field other than the detailed modifications in the sequence and arrangement of elementary functions. We must keep open the view that many distinctive psychological operations take form only within a social field and that the changes they produce alter individuals at their center."[120]

If we now consider the twentieth century development in American anthropology that has been most familiarly labelled as the study of personality and culture, we are struck by a very interesting historical fact. Whereas other psychological influences upon anthropology all can be derived from psychology in the somewhat narrow academic sense, in this case the source lay entirely outside this tradition. The interest that led to investigations in the personality and culture area, in short, represents but a single aspect of the tremendous influence which psychoanalytic theories have had upon the whole intellectual climate of our time, especially those disciplines concerned with human behavior, including, of course, academic

[120] Asch, *op. cit.*, copyright 1953 by Prentice-Hall, Inc., New York, pp. 76, 78.

psychology itself.[121] These theories were novel and they were radical; they challenged fundamental and highly cherished assumptions about the nature of man; it was inevitable that they should arouse antagonism and be subjected to the sharpest criticism. But quite aside from any technical detail, a few significant general facts may be noted here which have a direct bearing upon the reaction of anthropologists to them.[122]

In the first place, it is interesting to note that these new theories grew out of psychotherapy with individual patients and that the actual life experiences of the patient became a focal point of attention. This kind of empirical material, quite aside from any interpretation of it in relation to the psychodynamics of individual adjustment, was a far cry from the kind of information the cultural anthropologist was accustomed to collect. His aim was to obtain data that would permit valid descriptive generalizations about the cultural attributes of human populations considered in their group aspects, rather than about the component individuals of such populations considered as functioning personalities. This is what led Sapir to observe that: "It is what all the individuals of a society have in common which is supposed to constitute the true subject matter of cultural anthropology and sociology. If the testimony of an individual is set down as such, as often happens in our anthropological mono-

---

[121] Appraising the situation from the standpoint of psychoanalysis, Heinz Hartmann, Ernst Kris, and Rudolph M. Loewenstein ("Some Psychoanalytic Comments on 'Culture and Personality' " in *Psychoanalysis and Culture: Essays in Honor of Géza Róheim,* New York: International Universities Press, 1951, pp. 3–4) write: "The juxtaposition of culture and personality designates a vast area of research, rich in promise and stimulation. Investigators working in this area tend to point with pride to their attempts at interdisciplinary cooperation. Their views, however, differ frequently and in essential points, and one may well gain the impression that we are faced with a transitory period, in which affairs are unsettled and results tentative." The authors go on to say that the point of departure for their comments is the fact "that many psychological propositions used by workers in the field are derived from psychoanalysis. During the present phase of historical development the scientific access to problems of human conflict, and hence of personality, initiated by Freud, is exercising influence on the social sciences which previously had relied on common-sense psychology [sic]. . . . No other branch of the social sciences has taken this call for new data more seriously than Anthropology. A wealth of material assembled during the best part of a generation is witness to the fruitfulness of the contact."

[122] Since Kluckhohn, *op. cit.,* has so ably surveyed the anthropological literature which reflects this influence in its earliest stages, the reader is referred to his article for documentation of the work of such pioneers as Sapir, Benedict, and Mead, as well as others.

graphs, it is not because of an interest in the individual himself as a matured and single organization of ideas but in his assumed typicality for the community as a whole."[123] Although P. W. Schmidt, as early as 1906, had called attention to the need for more information about individuals in non-literate cultures,[124] and Radin had published the "Personal Reminiscences of a Winnebago Indian" in 1913,[125] the personal document approach in anthropology[126] was greatly stimulated and the qualitative aspects of the material collected assumed more clinical overtones once its psychological significance became apparent. This emphasis was promoted, for instance, by Edward Sapir, who familiarized himself with psychoanalytic theory at an early date.[127] Kluckhohn remarks that "although his own published contributions to this field are relatively slight, the influence of Edward Sapir in the direction of arousing interest in all kinds of personal documents is enormous."[128] Goldenweiser (1940) discussing recent trends in anthropology chooses "The Individual" as one of his three major headings and remarks that "among the anthropologists of the new functionalism with their disciples and followers, and among several of the younger members of the Boas school, the search for the individual has become something of an obsession."[129]

In the second place, the individuals from whom the psychoanalysts derived their basic data were those whose life adjustments were

[123] Edward Sapir, "Cultural Anthropology and Psychiatry," *Journal of Abnormal and Social Psychology*, Vol. 27, 1932, pp. 229–242.

[124] P. W. Schmidt, "Die Moderne Ethnologie," *Anthropos*, Vol. 1, 1906, pp. 592–644.

[125] Paul Radin, *Journal of American Folklore*, Vol. 26, 1913, pp. 293–318.

[126] See Clyde Kluckhohn, "The Personal Document in Anthropological Science" in Louis Gottschalk, Clyde Kluckhohn, and Robert Angell in *The Use of Personal Documents in History, Anthropology, and Sociology*, New York: Social Science Research Council, Bulletin 53, 1945, pp. 79–173; and A. L. Kroeber, "The Use of Autobiographical Evidence" in *The Nature of Culture*.

[127] His reviews of Freud's *Delusion and Dream*, and Oskar Pfister's *The Psychoanalytic Method* appeared in *The Dial*, the same year (1917) that he took issue with Kroeber in the *American Anthropologist* ("Do We Need a Superorganic?"). Kroeber (*The Nature of Culture*, copyright 1952 by the University of Chicago, p. 300) has reported the fact that he practiced psychoanalysis in San Francisco (1920–1923) but found that this experience gave him no insights that helped him understand culture better. Any negativism towards "culture-personality," he says is the "result of disillusionment rather than of prejudgment."

[128] Kluckhohn, *The Personal Document*, p. 88.

[129] "Leading Contributions of Anthropology to Social Theory," *Contemporary Social Theory*, Barnes, Becker, and Becker, eds., p. 480.

disturbed in one way or another. Thus the orientation of the analyst was, to begin with, directed towards the "abnormal," rather than the "normal." This orientation likewise was in contrast to the approach of the anthropologist. But this very orientation itself appears to have attracted the attention of anthropologists because it raised a question relevant to the known diversity in culturally constituted norms and value systems. Wide variation in social organization, economic life, art forms, and morals was a familiar fact. Might not psychopathological phenomena, at the level of individual behavior, be considered in a relativistic frame of reference? A number of articles began to appear that dealt with this general subject, including some case material. In the same year (1934) that Benedict's *Patterns of Culture* was published, her article "Anthropology and the Abnormal" appeared.[130] About the same time John M. Cooper published an article entitled "Mental Disease Situations in Certain Cultures: A New Field of Research."[131] Hallowell likewise discussed this subject[132] and Kroeber contributed a chapter to a book on *The Problem of Mental Disorder*.[133] These all followed the seminal article "Cultural Anthropology and Psychiatry" by Edward Sapir that had appeared in 1932.

The most significant aspect, however, of psychoanalytic theory that directly influenced anthropologists was the novel structural concepts of the human personality it offered. Not only was the personality conceived as a functioning whole; psychoanalytic theories dealt with the genesis, motivational patterns, and dynamics of personality. Neither academic psychologists nor orthodox psychiatrists had developed anything comparable. As Gillin has pointed out "the word 'personality' is not mentioned in the index, published in 1928, of the first forty volumes of the *American Anthropologist*. . . . The index of the publications of the Bureau of American Ethnology is also barren of the word, as is also that of Lowie's *The History of Ethnological Theory*. . . ."[134] And Kroeber remarks that "as late as 1915 the very word 'personality' still carried overtones chiefly of

[130] *Journal of General Psychology*, Vol. 10, 1934, pp. 59–80.

[131] *Journal of Abnormal and Social Psychology*, Vol. 29, 1934–1935, pp. 10–17. See also "The Cree Witiko Psychosis," *Primitive Man*, Vol. 6, 1933, pp. 20–24.

[132] A. I. Hallowell, "Culture and Mental Disorder," *Journal of Abnormal and Social Psychology*, Vol. 29, 1934–1935, pp. 1–9.

[133] Edited by Madison Bentley, New York: 1934.

[134] Gillin, *op. cit.*, 1939, p. 681.

piquancy, unpredictability, intellectual daring: a man's personality
was much like a woman's 'it'."[135] Anthropologists had not only bypassed the individual, normal or abnormal; concentration upon abstracted culture "traits" and "complexes" and references to individuals as "carriers" of culture even implied some basic dichotomy
between man and culture. A vital issue was obscured. How was it
possible for societies with all the widely different patterns of culture
with which the anthropologist was becoming more and more familiar
to function as integral units, if there were not some intimate connection between the psychological structure of their component individuals and the systems of culture they exhibited? Did not the very
existence of a *human* social order raise the question of personality
organization? Such questions could scarcely be attacked until some
working hypothesis about the nature of the human personality as a
structural whole had been developed. The older theories of the nature of the human mind were not adequate.

While anthropologists reacted negatively to Freud's own interpretations of the psychological significance of certain cultural phenomena, as exemplified in *Totem and Taboo*,[136] it soon dawned upon
them that there were implications in his theory that he himself did
not clearly envisage. The positive point which struck fire was the
hypothesis that early childhood training and experiences were of
paramount importance with relation to the kind of personality
structure developed by the individual; that parents or others who
were responsible for child training were surrogates of the larger cultural whole, and that the same socialization process that anthropologists had always assumed to be necessary for the transmission of
culture from generation to generation was, at the same time, the
process in which the personality of the individual was structured.
Here again, the cross-cultural data suggested that customary modes
of child training and the relations of parents and children were
variable features in human societies. *Group* differences in child
training might correlate with a typical personality structure on the
one hand and a characteristic culture pattern on the other. And since
psychoanalytic theory emphasized the crucial importance of certain kinds of infant experience, such as bowel training and weaning,
it was possible to obtain information about such facts through field
investigation.

   [135] *The Nature of Culture*, copyright 1952 by the University of Chicago, p.
116.
   [136] See, e.g., Kroeber, *op. cit.*, for the reviews he wrote in 1920 and in 1939.

Besides this, psychoanalytic theory presented anthropologists, as well as psychologists, with a novel concept of human nature. In anthropology the "psychic unity" of man had become a vague concept and it had been approached chiefly with questions of racial differences in mind. It had no reference to any constructs that could be directly related to the dynamics of the personality structure of man. By documenting the role of unconscious, biologically rooted impulses in man's conduct, and the necessity for some compromise with the demands imposed by an organized, culturally constituted mode of social existence, psychoanalytic theory had transcultural implications. Ubiquitous psychological mechanisms such as conflict, repression, sublimation, rationalization, etc., were said to be generic to the adjustment of all human beings, whatever the culture patterns of their society. Sapir grasped the importance of this contribution of psychoanalytic theory at once. In 1921 he expressed the opinion that "the really valuable contribution of the Freudian school seems to me to lie in the domain of pure psychology. Nearly everything that is specific in Freudian theory, such as the 'Oedipus-complex' as a normative image or the definite interpretation of certain symbols or the distinctively sexual nature of certain infantile reactions, may well prove to be either ill-founded or seen in distorted perspective, but there can be little doubt of the immense service that Dr. Freud has rendered psychology in his revelation of typical psychic mechanisms. . . . Psychology will not willingly let go of these and still other Freudian concepts, but will build upon them, gradually coming to see them in their wider significance."[137] Sapir proved entirely correct in this judgment. For as Hilgard points out (1949): "The mechanisms of adjustment were the features of Freudian theory that we earliest domesticated within American academic psychology. They now have a respectable place in our textbooks, regardless of the theoretical biases of our text-book writers."[138] It can be assumed, I think, that they have become an integral part of anthropological thinking about human behavior as well.

The history of culture and personality studies in anthropology, consists largely of the way in which psychoanalytic concepts, constructs, and theories have been accepted in whole or in part,

[137] In a review of Rivers, *Instinct and the Unconscious,* reprinted in *Selected Writings of Edward Sapir,* edited by D. Mandelbaum, Berkeley: University of California Press, 1949.

[138] Ernest R. Hilgard, "Human Motives and the Concept of the Self," *The American Psychologist,* Vol. 4, 1949, p. 374 (Presidential Address, American Psychological Association, 1949).

modified or combined with learning theory, and related to cultural or sociological material. What is particularly impressive is that Freud rather than Jung, Rank, Adler, or other analysts has proved to be the most influential figure, although Kluckhohn draws attention to the fact that Radin, in the late twenties, predicted that Jung would have a greater influence than the others.[139] In his recent discussion of *Jung's Psychology and Its Social Meaning*, Progoff stresses the fact that "In the United States, Radin's work stands out as the major exception to current academic anthropological views and the general lack of appreciation for Jung's concepts."[140] On the other hand, Radin appears to entertain critical reservations of the most fundamental

[139] Kluckhohn's reference is to "History of Ethnological Theories," *American Anthropologist*, 31, 1929, pp. 26–30, where Radin says: "It is the application of the psycho-analytical theories of Jung that is most likely to have the most profound influence upon ethnology. This is, after all, to be expected, if for no other reason than that Jung's attitude, in addition to containing entirely new concepts introduced by him, represents in a manner, the synthesis of current theories of psycho-analysis." Cf. *Social Anthropology*, New York, 1932, p. 16. In a journal promoted by Radin in the 1920's in which the "functional, dynamic, and human side of primitive culture" was to be particularly stressed, along with the psychological, the approach to Winnebago Myth Cycles employed was explicitly attributed to Jung. (*Primitive Culture. An International Journal Devoted to the Study of Social Anthropology.* Only a single issue, dated July 1926, appeared.) A four-fold chronological sequence was suggested which corresponds, says Radin, "to what I take to be the substance of Jung's theory. The Trickster period would then represent his undifferentiated, the Hare his imperfectly differentiated, the Red-Horn his well-differentiated, and the Twin his integrated libido" (pp. 12–13). Only the first cycle (the Trickster) was published in *Primitive Culture*. In a later publication which contains all four cycles (*Winnebago Hero Cycles: A Study in Aboriginal Literature*, Supplement to *International Journal of American Linguistics*, Vol. 14, No. 3, Memoir I, July 1948) Radin does not explicitly refer to the earlier publication of the Trickster Cycle in *Primitive Culture*. Nevertheless, he reiterates the fact that "these four cycles, within limits, lend themselves to a definite temporal sequence" (pp. 8–9) without, however, making their Jungian derivation so specific. It may be noted that Radin has contributed articles to *Eranos Jarhbuch*, a publication chiefly devoted to the application of various aspects of Jung's theories.

Kluckhohn likewise makes reference to William Morgan, an American psychiatrist influenced by Jung, who has published on the Navajo. But certainly no American anthropologist of eminence, except Radin, seems to have come under Jung's influence in any respect, although both Kroeber and Sapir reviewed his work at an early date, as well as that of Freud (Kroeber, "Analytical Psychology and Psychology of the Unconscious" in the *American Anthropologist*, 20 (1918); Sapir, "Psychological Types" in *The Freeman*, 8 (1923).

[140] Ira Progoff, *Jung's Psychology and Its Social Meaning*. An introductory statement of C. G. Jung's psychological theories and a first interpretation of their significance for the social science. (New York: The Julian Press, 1953, p. 274.)

nature. "If, for the early evolution of man's culture and thought," he says, anthropologists "cannot accept the findings and conclusions of Freud's psychoanalysis or Jung's complex psychology, they must at least do more than simply state their objections, however justified these appear. They owe it to their subject to attempt to utilize as many as possible of the suggestions Freud, Jung, and their followers have thrown out and the new lines of inquiry they have initiated, this, despite the fact that they may, as I personally do, disagree fundamentally with the viewpoint, the methods, and the conclusions of the latter."[141]

In England, Seligman applied Jung's psychology of types to both "savage" and "civilized" peoples in his Presidential Address to the Royal Anthropological Institute in 1924. Subsequently, John Layard, although not a professional anthropologist in the strictest sense of the term, but whose *Stone Men of Malekula*, 1942, reports the field work he did in the New Hebrides many years before, has made systematic use of Jungian psychology. His interpretations of the significance of the incest tabu, marriage sections in Australia and Ambrym, myths, ceremonies, and other cultural data in these terms lead to highly novel results.[142]

III. Current Developments and Future Trends

While the importance of psychoanalytic theory as the major stimulus which promoted the most striking and novel psychological inquiries and hypotheses developed by twentieth century anthropology should not be minimized, I venture to say that this specific impetus already has passed its peak. So far as the future relations of

---

[141] Paul Radin, *Winnebago Hero Cycles, op. cit.*

[142] See in particular "The Incest Taboo and the Virgin Archetype," *Eranos-Jahrbuch*, 12 (1945), pp. 254–307 (this volume was a *Festgabe* for Jung on the occasion of his 70th birthday); "Primitive Kinship as Mirrored in the Psychological Structure of Modern Man," *British Journal of Medical Psychology*, Vol. 20 (1944–1946), pp. 118–134; "The Making of Man in Malekula," *Eranos-Jahrbuch*, 16 (1949), pp. 209–283; "Der Mythos der Totenfahrt auf Malekula," *ibid.* (1937); "Maze Dances and the Ritual of the Labyrinth in Malekula," *Folk-Lore*, 47 (1936). Layard is likewise the author of *The Lady of the Hare* (London: Faber and Faber, 1944). This is a study of the healing power of dreams, in which the author presents a detailed documentary account of an analysis of a woman which he made in which the hare is a central image; the latter part of the book is a comparative study of the mythology of the hare on a world-wide basis.

anthropology and psychology are concerned in the development of a science of social man the ground is already prepared for what may prove to be further influences emanating from psychology rather than from psychoanalytic psychiatry. Furthermore, it seems likely that such influences will turn out to be much more positive and fructifying than those exerted at any period in the past for several reasons. First, anthropologists are rapidly developing a better psychological perspective relative to their own data; secondly, certain lines of specialized inquiry in psychology are more directly related to the interests of anthropologists than ever before; and, in the third place, psychologists are becoming more acutely aware of the data of anthropology and their relation to their own inquiries. There are a number of focal areas in which developments are now taking place which are of direct concern to anthropologists.

1. Personality Theory and Personality Tests. Personality theory is now being rapidly assimilated into general psychology. In a recent chapter dealing with this area of inquiry MacKinnon and Maslow conclude their historical review as follows:[143]

> The field of personality is extraordinarily rich in fascinating hunches, hypotheses, and theories. These have come in greatest number and with greatest richness from psychiatrists and clinicians, who, however, have not been noted as a group for experimental emphasis or methodological sophistication. Fortunately, there are mounting signs of an integration of personality theory into the general theory of psychology. To the extent that a true rapprochement is achieved, psychology will increasingly focus its attention upon the more dynamic and more molar aspects of behavior, and in turn, the various and often conflicting theories of personality development, structure, and function will be subjected to the test of rigorous scientific investigation which alone can yield the facts required for their ultimate rejection, modification, or validation.

Anthropologists who are interested in personality and culture will have to keep abreast of such developments in psychology and it may be that psychiatry will be vitally influenced by anthropology and psychology rather than the reverse. It has been said already that "psychiatry rests upon three main disciplines, medicine, psychology,

---

[143] Donald W. MacKinnon and A. H. Maslow, "Personality," Chapter 13, Harry Helson, ed., in *Theoretical Foundations of Psychology*, New York: Van Nostrand, 1951. See also, Donald W. MacKinnon, "Fact and Fancy in Personality Research" in *The American Psychologist*, Vol. 8, 1953, pp. 138–146, and Robert Leeper, "Current Trends in Theories of Personality" in *Current Trends in Psychological Theory*, Pittsburgh: U. of Pittsburgh Press, 1951.

and social anthropology. In this framework psychology is conceived of as dealing broadly with behavior, but more particularly with psychodynamics. Psychodynamics is mainly now a clinical and speculative subject. The challenge to psychology is to help put it on a scientific basis. There is little evidence that the job can be done alone by psychiatry."[144] A specific example which may be indicative of a general trend is the position taken by Whiting and Child in their recent study of child training and personality. Their investigation is "oriented toward testing generalized hypotheses applicable to any case" rather than "toward seeking concrete understanding of specific cases. . . . Most studies which have been concerned with general hypotheses about culture and personality," they write, "have tended to express those hypotheses in terms of the concepts of psychoanalytic theory. Here we differ sharply from our predecessors. We have preferred to use, to a much greater extent, concepts drawn from the general behavior theory that has been developed by academic and experimental psychologists."[145]

A somewhat analogous situation is emerging in the case of personality tests. Those of the projective type, of which the Rorschach and the Thematic Apperceptive Test are representative and most widely used, came out of the clinic rather than the psychological laboratory. Herman Rorschach was a Swiss psychiatrist; Henry A. Murray, the inventor of the TAT has a psychiatric as well as a psychological background. Both tests have been used by psychiatrists and clinical psychologists, but it is psychologists who are now undertaking the task of appraising their validity and reliability in a rigorous fashion. American anthropologists first began to use the Rorschach test in their studies of personality and culture in the late 1930's and, somewhat later, the TAT.[146] Much of the work done so

[144] Paul E. Huston, "Some Observations on the Orientation of Clinical Psychology," *The American Psychologist*, Vol. 8, 1953, p. 196. See likewise *Psychiatry and Medical Education*, Report of the 1951 Conference held at Cornell University, Ithaca, N.Y., June 21–27, 1951, Organized and Conducted by the American Psychiatric Association and the Association of American Medical Colleges, Washington: American Psychiatric Association, 1952, Chapter 5, "Human Ecology and Personality in the Training of Physicians."

[145] Whiting and Child, *op. cit.*, pp. 5 and 13.

[146] See A. Irving Hallowell, "The Rorschach Technique in the Study of Personality and Culture," *American Anthropologist*, Vol. 47, 1945, pp. 195–210, and Jules Henry and Melford E. Spiro, "Psychological Techniques: Projective Tests in Field Work" in A. L. Kroeber, ed., *Anthropology Today, An Encyclopedic Inventory*, Chicago: University of Chicago Press, 1953.

far has been of a pioneer nature, carried on during a period when even clinical psychologists made less use of the Rorschach than today. Elsewhere[147] I have discussed this technique in relation to personality and culture studies, as well as to certain broad theoretical issues that concern the nature and condition of human perception, based upon an overall appraisal of the protocols we now have from hundreds of individuals with varying cultural backgrounds in different parts of the world. In the conclusion to this paper, I have stressed the following points: (a) the importance of adequate training for those who propose to use the test; (b) the necessity for carefully evaluating its possibilities, as well as its limitations, in relation to the aims of a systematically designed study; (c) the presentation of the results obtained in accordance with the same requirements demanded of those who use the Rorschach in clinical practice; (d) and that such results be subjected to the same standards of evaluation. In order that such standards be met a closer cooperation will be necessary with psychologists who are using the test as well as with those who are appraising it. At the same time, the results obtained from the use of the Rorschach by anthropologists should be greatly enhanced.

2. **Learning Theory.** Learning theory is one of the most highly specialized areas of recent psychological research, especially in the United States. Developments in this field cannot be neglected in the future by the psychologically oriented anthropologist who is interested in a science of social man. Learning theory is as relevant to a deeper understanding of the nature of culture viewed in terms of the remarkable persistence of characteristic patterns in time and the processes of culture change and acculturation, as it is to an understanding of the psychological aspects of the socialization process and the acquisition of a personality structure by the individual. Nevertheless, within psychoanalytic theory there is very little attention paid to learning. What has been stressed, on the contrary, is a succession of developmental stages closely correlated with maturational processes. However firm the evidence for any such stages may turn out to be,[148] a ubiquitous succession of this sort does not help us

[147] A. Irving Hallowell, "The Rorschach Test in Personality and Culture Studies" in Bruno Klopfer, ed., *Further Contributions to the Rorschach Technique*, Vol. II, *Applications*. (In press.)

[148] In cooperation with members of the Gesell Institute of Child Development, Margaret Mead and Frances Cooke Macgregor (*Growth and Culture. A*

to understand how differential personality characteristics, which have group as well as idiosyncratic dimensions, are acquired, or the psychological consequences of what is learned in one society as compared with another. Consequently, Murdock[149] has called for an integration, at the level of theory, of contributions from psychoanalysis and psychiatry relating to personality development, of contributions from behavioristic psychology to learning theory, combined with the contributions of sociology and cultural anthropology. And, at the beginning of their exposition of "A Dynamic Theory of Personality" Mower and Kluckhohn explicitly state that they have drawn upon "three relatively independent lines of scientific development: psychoanalysis, social anthropology, and the psychology of learning."[150] Personality theory and learning theory cannot be divorced from each other in a developing science of social man because the persistent attributes of a socio-cultural system are dependent upon psychological processes. The very existence of a culture is dependent upon the fact that certain aspects of it become an integral part of the functioning individual. Tolman has remarked that "psychology is in large part a study of the internalization of society and of culture within the individual human actor,"[151] and Newcomb speaks of the individual as having "somehow got society inside himself. Its ways of doing things become his own."[152] If this were not so no one could live his culture nor could he hand it on. A science of social man demands reliable knowledge of the learning process; the recognition of the bare fact that learning occurs or *ad hoc* references to this

---

Photographic Study of Balinese Childhood, Based upon Photographs by Gregory Bateson, Analyzed in Gesell Categories, New York: G. P. Putnam Sons, 1951) have attacked this problem and presented results which show certain broad similarities as well as significant differences in developmental patterns.

[149] George P. Murdock, "The Science of Human Learning, Society, Culture, and Personality," *Scientific Monthly*, Vol. 69, 1949, pp. 377–381.

[150] O. H. Mower and Clyde Kluckhohn, "Dynamic Theory of Personality" in *Personality and the Behavior Disorders*, edited by J. McV. Hunt, New York: Ronald Press, 1944, Vol. 1, pp. 69–135. The learning theory used is of the S-R type. It is stated (p. 79) that "the great unifying principle in the version of learning theory that is here espoused is the proposition that all behavior is *motivated* and that all learning involves reward."

[151] E. C. Tolman in Talcott Parsons and Edward A. Shils, eds., *Toward A General Theory of Action*, Cambridge: Harvard University Press, 1951, p. 359.

[152] Theodore M. Newcomb, *Social Psychology*, New York: Dryden, 1950, p. 6. See likewise Melford E. Spiro, "Culture and Personality: The Natural History of a False Dichotomy," *Psychiatry*, Vol. 14, 1951, pp. 19–46, and Talcott Parsons, "The Superego and the Theory of Social Systems," *ibid.*, Vol. 15, 1952, pp. 15–25.

process in connection with certain problems can only lead to the oversimplification or distortion of vital questions.

Many years ago (1927) A. A. Goldenweiser made some observations on the "Psychological Postulates of Evolution" during the course of which he attempted to draw an analogy between the process of cultural change and the learning process. His hypothesis was that since cultural change involves the overcoming of the inertia of past habits there is a period of delay during which cumulative pressures build up ("or psychologically, summation of stimuli") followed by the overcoming of resistance so that "the change comes—with a spurt." He then goes on to say that "we know from the study of the learning process that it is not gradual but jerky. So also with culture, for, from one angle, culture is learning and the psychology also is the same. The delay comes from inertia due to pre-existing habits, only that in the case of culture the inertia of the individual is greatly reinforced by institutional inertia. This lengthens the delay and adds to the explosive character of the change when it does come."[153] Surely, a little knowledge is a dangerous thing. While learning theory is relevant to the empirical facts of culture change, the analogical approach of Goldenweiser is misleading and irrelevant.

Although the need for reliable knowledge about the learning process is generally recognized, nevertheless the fact remains, as Newman says, that "there is no single theory of learning. Perhaps there cannot be." At the same time there is no lack of learning *theories* and these theories reflect, as in a mirror, various psychological orientations. "Ask a psychologist," continues Newman, "what is his view of learning and you will discover his scientific credo, his beliefs and prejudices, his love of empirical fact, of deductive elegance, of operational rigor, of didactic power."[154] The uses of learning theory already made by some anthropologists reflect this very fact. So far, the

[153] Goldenweiser, "Anthropology and Psychology," reprinted in *History, Psychology and Culture*, copyright by Alfred A. Knopf, Inc. p. 74.

[154] Edwin B. Newman, "Learning" in *Theoretical Foundations of Psychology*, New York: Van Nostrand, 1951, p. 390. For other surveys and discussions of learning theory see, in particular, Ernest R. Hilgard, *Theories of Learning*, New York: Appleton-Century, 1948; James J. Gibson, "The Implications of Learning Theory for Social Psychology" in M. J. G. Miller, ed., *Experiments in Social Process: A Symposium on Social Psychology*, New York: McGraw-Hill, 1950; and for current trends in "perceptual learning" theory, Harry F. Harlow, "Learning Theories," Wayne Dennis, Robert Leeper, et al., in *Current Trends in Psychological Theory*, Pittsburgh: University of Pittsburgh Press, 1951.

behavioristic theory of Hull has furnished the major model.[155] Gillin has made use of it in his text, *The Ways of Men.* John Whiting uses this model of learning theory systematically in his analysis of the acquisition of Kwoma culture by the child, and Beatrice Whiting in her analysis of patterns of social control.[156] Miller and Dollard's book, *Social Learning and Imitation,* representing the same point of view, contains a chapter on "Copying in the Diffusion of Culture." "Copying can become an acquired drive," they write, "providing copying behavior has become rewarded. Conditions of social contact offer the best opportunity for rapid copying, since they bring the learner into contact with the model and critic who can rapidly elicit the correct response. The prestige of the model is the crucial matter in mobilizing the copying drive and setting in motion the attempt to match responses. Copying is rarely exact, owing to various circumstances, chief among them the pressure put on the incoming habit by the preëxisting matrix of the receiving culture."[157] Gillin[158] and Hallowell[159] have made use of the Miller-Dollard paradigm of learning in relation to acculturation, Horton[160] in his cross-cultural study of alcoholism, and Holmberg[161] in his analysis of the effects of the frustration of the hunger drive among the Siriono. Spiro, on the

[155] The reason particular anthropologists have turned to S-R learning theory seems to have been largely circumstantial. For, in fact, both Tolman and Lewin were concerned with social learning and applied their theories in this area, but up to the present time they have not exercised any discernible influence on anthropologists.

[156] John W. M. Whiting, *Becoming a Kwoma: Teaching and Learning in a New Guinea Tribe,* New Haven: Yale University Press, 1941; Beatrice B. Whiting, "Paiute Sorcery," *Viking Fund Publications in Anthropology,* No. 15, New York: Wenner-Gren Foundation, 1950.

[157] Neal E. Miller and John Dollard, *Social Learning and Imitation,* New Haven: Yale University Press, 1941. In their preface the authors state their obligation to "Prof. George P. Murdock for aid in preparing the chapter on diffusion and for a most helpful reading of the manuscript as a whole. Prof. B. Malinowski gave us a single, but significant, interview in the diffusion problem."

[158] John Gillin, "Parallel Cultures and the Inhibitions to Acculturation in a Guatemalan Community," *Social Forces,* March, 1945; "Acquired Drives in Culture Contact," *American Anthropologist,* Vol. 44, 1942, pp. 545–554.

[159] A. Irving Hallowell, "Sociopsychological Aspects of Acculturation" in *The Science of Man in the World Crisis,* Ralph Linton, ed., New York: Columbia University Press, 1945.

[160] Donald Horton, "The Functions of Alcohol in Primitive Societies: A Cross-Cultural Study," *Quarterly Journal of Studies on Alcohol,* Vol. 4, 1943, pp. 199–320.

[161] Allan R. Holmberg, *Nomads of the Long Bow. The Siriono of Eastern Bolivia,* Smithsonian Institution, Institute of Social Anthropology, Pub. No. 10, Washington: United States Government Printing Office, 1950.

other hand, turning to learning theories for assistance in explaining
the process by which the Ifaluk acquire a belief in malevolent ghosts,
expresses the opinion that "social scientists may have been too hasty
in accepting the experimental laws of learning because some cultural
phenomena, at least, cannot be explained by current learning
theories."[162] And in Gibson's opinion the experimentally founded
theories, such as Hull's, "do *not* fit the facts of social learning in one
important respect. As now formulated, they do not account for the
astonishing prevalence of moral behavior among human adults." The
need for a theory of *social* learning is, therefore, acute and Gibson
warns that, "If the social psychologist does not formulate a theory
of learning, the cultural anthropologist will have to do so—and
also the psychiatrist, the clinician, the educator, and the student of
child development."[163]

One anthropologist, Gregory Bateson,[164] has made an important
contribution to learning theory. Harry F. Harlow, an exponent of
perceptual learning theory writes: "After we had published a formulation of our theory we found that it had in principle already been
proposed by the anthropologist Bateson, who had described the
phenomenon as 'deutero-learning' and had italicized the words which
we believed most descriptive of the theory, the animal *learns to
learn*. Our contribution has been to provide extensive, rigidly controlled experimental data in support of such a theoretical formulation. Bateson wrote: 'We need some systematic framework or classification which shall show how each of these habits is related to the
others, and such a classification might provide us with something
approaching the chart we lack.' Our own researches, we believe,
provide or have begun to provide this 'systematic framework.' Bate-

[162] Melford E. Spiro, "Ghosts: An Anthropological Inquiry into Learning
and Perception," *Journal of Abnormal and Social Psychology*, Vol. 48, 1953,
pp. 376–382.

[163] Gibson, *op. cit.*, pp. 151 and 152. As examples of collaborative efforts he
cites Mowrer and Kluckhohn, Miller and Dollard, and the report of a committee
of the National Society for the Study of Education, 1942. For an appraisal of
learning theory with special reference to the concept of culture used by Miller
and Dollard, see Omar K. Moore and Donald J. Lewis, "Learning Theory and
Culture," *Psychological Review*, Vol. 59, 1952, pp. 380–388. Moore is a sociologist, Lewis a psychologist.

[164] Gregory Bateson, "Social Planning and the Concept of Deutero-learning"
in *Science, Philosophy and Religion*, Second Symposium, New York: Conference on Science, Philosophy and Religion, 1942, pp. 81–97. Reprinted in
abridged form in T. H. Newcomb and E. L. Hartley, eds., *Readings in Social
Psychology*, New York: Holt, 1947, pp. 121–128.

son in his paper discusses the anthropological implications of such a theory. We have indicated its possible role in personality formation in a theoretical paper."[165] Mutual influences of this sort indicate, in principle, the possible benefits for a developing science of social man that can be derived from the coordination of knowledge in psychology and anthropology in a single problem area.

3. **Social Psychology.** The tardy development of a theoretically mature and experimentally oriented social psychology has handicapped the relations between anthropology and psychology in the past. "The rapid growth and transformation of social psychology," says Brewster Smith, "has come only in the recent postwar years."[166] Just as in the case of learning theory, there is no united front in social psychology. But it is perhaps significant that social psychologies written from a Gestalt point of view have commanded more and more attention. Krech and Crutchfield have argued recently, moreover, that general psychology *is* social psychology:

> . . . not only do the same fundamental principles apply to both social and non-social behavior, but the general psychologist as well as the social psychologist is literally forced to study the behavior of man as a social being. Whether we are studying the behavior of man in a laboratory, in the clinic, or in a crowd, whether we are studying his perception of colored papers, his performance on an intelligence test, or his decision about participating in a lynching, we are studying the behavior of a man as influenced by his past and present interpersonal relationships reaching into each of his psychological activities no matter how simple or apparently remote. As a consequence every man lives in a social world, and no psychologist, whatever his interests, does or can study the behavior of an asocial man.[167]

Whatever the reaction to such a statement on the part of psychologists may be, it is thoroughly intelligible, and even acceptable, to many anthropologists. It is the kind of position that could become a common meeting ground. At the same time it is a far cry from the

---

[165] Harry F. Harlow, "Levels of Integration Along the Phylogenetic Scale. Learning Aspect" in *Social Psychology at the Crossroads*, edited by J. H. Rohrer and M. Sherif, New York: Harper and Brothers, 1951, pp. 138–139, and "The Formation of Learning Sets" *Psychological Review*, Vol. 56, 1949, pp. 51–65. For comments on Harlow and with references to culture see also O. Hobart Mowrer, *Learning Theory and Personality Dynamics*, New York: Ronald Press, 1950, pp. 332 ff.

[166] M. Brewster Smith, Review, "Some Recent Texts in Social Psychology," *Psychological Bulletin*, Vol. 50, 1953, p. 150.

[167] By permission from *Theory and Problems of Social Psychology*, by David Krech and Richard S. Crutchfield, copyright 1948, McGraw-Hill Book Company, Inc., pp. 7–8.

position which Kroeber, Lowie, and Wissler identified as character-istic of psychologists. But it is close to that of Bartlett, whom Krech regards as a pioneer[168] and, farther back, we cannot but recall the implications of the position taken by the Herbartians—Waitz and Bastian. They, too, would probably have accepted this fundamental frame of reference, as would Rivers, Seligman, or Boas.

The major implication of this psychological orientation in con-trast to that of the traditional experimental psychologist, is that whereas the latter, as Krech points out, "has always assumed, usually implicitly, that in studying perception, for example, he was study-ing a process independent of the social background, affiliations, prejudices, and values of the receiver,"—and that "among the vari-ables which the perception psychologist, or the learning psycholo-gist included in this theoretical model of the perceiver and learner there was no room for social values and mores," the former assumes that "every basic psychological process of man—his perception, learning, remembering, wanting, feeling, etc.—must be understood within the social context of the perceiver, the learner, the remem-berer, the wanting man."[169] Although Krech does not explicitly say so, it is obvious that the cross-cultural data of the anthropologist are directly relevant to the investigation of such processes. Not only the personality structure but the complex relations that exist be-tween such processes as those mentioned and the psychological field of behavior as structured by the cultural variables that characterize different socio-cultural systems, suggest special problems for in-quiry.

A number of psychologists have become interested in "social perception," and the relation of perception and personality.[170] "How-

[168] David Krech, "Psychological Theory and Social Psychology" in Harry Nelson, ed., *Theoretical Foundations of Psychology*, New York: Van Nostrand, 1951, p. 666. Referring to Bartlett's *Remembering* (1932) Krech says: "Here, for the first time, was the principle generalized that while there appeared to be universal laws of memory and perception, the nature of these processes and the specific errors in memory and perception which people made could not be understood except in terms of the cultural backgrounds and socially determined interests and mores of the subjects."

[169] Krech, *ibid.*, p. 665.

[170] See J. S. Bruner and Leo Postman, "An Approach to Social Perception" in *Current Trends in Social Psychology*, Pittsburgh: University of Pittsburgh Press, 1948; Robert R. Blake and Glenn V. Ramsey, eds., *Perception: An Ap-proach to Personality*, New York: Ronald Press, 1951. Wayne Dennis has a chapter in this volume dealing with "Cultural and Developmental Factors in Perception."

ever varied the terminology," writes Gibson, "a central conviction is shared by these students of social perception: they are sure that the doctrine of the passive perceiver who simply mirrors the world is a myth and is now disposed of for good. What a man perceives, they say, depends on his personality and his culture. Men of different cultures perceive quite different worlds. There are, in short, folkways of perception." But Gibson advises caution in drawing any categorical conclusion without qualification. He goes on to say that "an enthusiasm for social psychology and a sense of the urgency of its problems is something that I share. But I cannot agree with the social perceptionists that the kind of problems they study are the prototype of all problems in perception. I would be willing to agree with them that all perception is in a certain sense socialized perception if they were willing to agree in turn that all perception is just as truly psychophysical perception."[171] However this issue may be resolved, it is of vital interest from an anthropological[172] as well as a psychological point of view and its solution is of fundamental importance to a science of social man.

4. **Human Nature.** Asch, in the first chapter of his *Social Psychology*, reaffirms the notion that "it is the goal of psychology to furnish a comprehensive doctrine of man, one that will provide a tested foundation for the social sciences." At the basis of these disciplines, he says, "there must be a *comprehensive conception of human nature*"[173] (italics ours). Almost twenty years before this volume appeared Paul Radin at the very end of his book *The Method and Theory of Ethnology*, expressed the view that the study of cultural anthropology ultimately led to the same central problem, but he was inclined to look to psychoanalysis for illumination. "Before concluding," he wrote, "let me refer briefly and superficially to what is perhaps the core of all investigations of culture: can we ever arrive at any satisfactory knowledge of what constitutes human nature? To say with Boas and so many ethnologists and sociologists that the culture picture hides this knowledge from us forever is a counsel of

[171] James J. Gibson, "Theories of Perception" in *Current Trends in Psychological Theory*, Pittsburgh: University of Pittsburgh Press, 1951, pp. 94–95, and *The Perception of the Visual World*, Boston: Houghton Mifflin, 1950.

[172] See, for example, Verne F. Ray, "Techniques and Problems in the Study of Human Color Perception," *Southwestern Journal of Anthropology*, Vol. 8, 1952, pp. 251–259; A. Irving Hallowell, "Cultural Factors in the Structuralization of Perception" in *Social Psychology at the Crossroads*.

[173] Asch, *op. cit.*, copyright 1953 by Prentice-Hall, Inc., New York, p. 5.

despair. Some significant light can surely be obtained, even if today the technique for this type of investigation has not as yet been perfected. Here again I feel that a psychoanalyst like Jung is on the correct trail."[174] More recently, Kroeber (1948) construing "human nature" as the "original" or "the general nature of man" views it as "a pretty vague and uncharacterized thing," perhaps a limiting factor in cultural development, but certainly not related causally to the manifold forms that culture assumes. He records his impression that, "psychologists have become very unwilling to discuss the inherent psychic nature of man. It is definitely unfashionable to do so. When the subject is faced at all, it is usually only to explain human nature away as fast as possible, and to pass on to less uneasy and more specific topics. Human nature is going the way the human 'mind' has gone. Instead, psychologists for the last few decades have increasingly dealt with the concept of personality."[175]

Although the term human nature has assumed a variety of meanings and so has become a weasel word, there are signs that the concept is far from moribund and that it may yet be given a more specific meaningful content which will prove to be of heuristic value in a science of social man. It is not likely, of course, that separately either psychologists or anthropologists will be able to formulate "a comprehensive conception of human nature." Besides, psychoanalytic theory is of great importance here. Roheim introduces his chapter on "The Unity of Mankind" with the quotation from Kroeber cited above and takes modern cultural anthropologists to task because of their negative attitude towards such unity.[176] Recently, however, there appears to be a developing interest in this topic on the anthropological side as represented, for example, in such publications as those of Bidney, Spiro, and Washburn.[177] Kluckhohn's

[174] Paul Radin, *The Method and Theory of Ethnology*, New York: McGraw-Hill, 1933, p. 267.

[175] A. L. Kroeber, *Anthropology*, New York: Harcourt, 1948, p. 619.

[176] Geza Roheim, *Psychoanalysis and Anthropology: Culture, Personality, and the Unconscious*, New York: International Universities Press, 1950. "The culturalist school will have nothing to do with a basic unity of mankind," he says (p. 489) and, "modern or 'cultural' anthropology tacitly negates the basic unity of mankind as well as the separateness of the individual—it sees only nations" (p. 491).

[177] David Bidney, "Human Nature and the Cultural Process," *American Anthropologist*, Vol. 49, 1947, pp. 375–399; Melford E. Spiro, "Human Nature in Its Psychological Dimensions," *ibid.* (in press); S. L. Washburn, "Evolution and Human Nature," *ibid.* (in press).

discussion of "Universal Categories of Culture" is focused in the same direction,[178] and Redfield has made some pertinent observations. Among other things he says: "In this territory we find ourselves characterizing human nature as a recognized and respectable subject matter of our science, and it is no surprise that certain branches of psychology and sociology have anticipated us. What is creditable to anthropology is that we learn from them. What is creditable to them is that they often modify their views under our influence. But conjointly this area of investigation is now opened. . . ."[179]

On the phylogenetic side, man's psychological distinctiveness from other primates, rather than any capacities or characteristics he may share with them now needs to be as closely examined and discriminated as does his morphological and physiological distinction from them.[180] Kenneth Oakley has observed that despite the fact that man "has been described as the reasoning animal, the religious animal, the talking animal, the tool-making animal, and so on, we are still in need of a working definition of man."[181] Any such definition, any answer to the query, What is man? involves the question, What is human nature? and in particular, the psychological distinctiveness of man. It is significant that LeGros Clark has expressed the opinion that "probably the differentiation of man from ape will ultimately have to rest on a *functional* rather than on an anatomical basis, the criterion of humanity being the ability to speak and to make tools"[182]

[178] Clyde Kluckhohn, "Universal Categories of Culture" in *Anthropology Today.*

[179] Robert Redfield in his discussion of "Culture, Personality and Society," the paper contributed by A. I. Hallowell to *Anthropology Today* in *An Appraisal of Anthropology Today,* edited by Sol Tax, Loren C. Eiseley, Irving Rouse, Carl F. Vogelin, Chicago: University of Chicago Press, 1953, p. 127. For a set of working assumptions regarding the nature of man used in applied social science see Alexander H. Leighton, *The Government of Men,* Princeton: Princeton University Press, 1945, Part II and Appendix; and *Human Relations in a Changing World: Observations on the Use of the Social Sciences* by the same author, New York: Dutton, 1949.

[180] See in particular, T. C. Schneirla, "Levels in the Psychological Capacities of Animals" in *Philosophy for the Future,* edited by Roy W. Sellars, V. J. McGill, Marvin Farber, New York: Macmillan, 1949; and Henry W. Nissen, "Phylogenetic Comparisons" in S. S. Stevens, ed., *Handbook of Experimental Psychology,* 1951.

[181] Kenneth Oakley, "A Definition of Man," *Science News,* 20, Penguin Books, 1951.

[182] W. E. LeGros Clark, *History of the Primates: An Introduction to the Study of Fossil Man,* 2nd ed., London: Handbook, British Museum, Department of Geology, 1950, p. 73.

(italics ours). But what underlies man's capacities, not only for speaking and tool-making, but the creation of manifold *varieties* of speech forms and technologies, world-views, value systems, and other forms of culture? What enables him to maintain them, elaborate them, change them and to readjust himself to influences derived from alien culture patterns?

It can hardly be maintained that man's peculiar and unique psychological characteristics are fully understood, despite the increasing depth and breadth of our knowledge derived from comparative psychology. It was only two decades ago that, writing as a mammalogist, G. S. Miller insisted that human sexuality was, for the most part, equivalent with primate sexuality.[183] Ford and Beach, combining data from psychology and cultural anthropology, give us a quite different and more reliable picture of human sexual behavior in phylogenetic perspective.[184] Hilgard, discussing learning theories, has pointed out that sometimes psychologists have given the impression that "there are no differences, except quantitative ones, between the learning of lower animals and primates, including man." He goes on to say, however, that "while this position is more often implied than asserted, it is strange that the opposite view is not more often made explicit—that at the human level there have emerged capacities for retaining, reorganizing, and foreseeing experiences which are not approached by the lower animals, including the other primates. No one has seriously proposed that animals can develop a set of ideals which regulate conduct around long-range plans, or that they can invent a mathematics to help them keep track of their experiences. Because a trained dog shows some manifestations of shame, or a chimpanzee some signs of cooperation, or a rat a dawning concept

[183] G. S. Miller, Jr., "The Primate Basis of Human Sexual Behavior," *Quarterly Review of Biology,* Vol. 6, 1931, pp. 379–410.

"The main characteristics of human sexual behavior, contrary to a widely prevalent belief, are not peculiar to man," the author says. There are only two human "specializations" which "have had any significance in directing the general course of cultural development." The first is "a stronger tendency than most other primates to form continuing sexual partnerships" (this was prior to Carpenter's observations on the monogamous gibbon). The second is "that in man alone of all mammals is the male known to be able to force his sexual will on an unconsenting or unconscious female, a peculiarity that seems to arise from human ingenuity combined with the human pelvic adjustments to the upright posture."

[184] Clellan S. Ford and Frank A. Beach, *Patterns of Sexual Behavior,* New York: Harper and Hoeber, 1951.

of triangularity, it does not follow that these lower organisms, clever as they are, have all the richness of human mental activity."[185]

Harlow, on the basis of animal experimentation, concludes that "the ability of the human animal to form and utilize multiple perceptual learning sets probably differentiates him from other animals to as great a degree as does the ability to utilize spoken language. Even monkeys and chimpanzees show very limited ability to form effective sets to differentiate tiny, detailed stimuli, or tiny, detailed variations between stimuli, in contrast to the human being whose incredible skill along these lines enables him to read the *Scientific American, Time,* the *Reader's Digest,* the *Daily Worker,* and the *Chicago Tribune.*"[186] Harlow does not point out that still other capacities of man enabled him to *invent* systems of writing, the alphabet, and the mechanical means that brought the printed page into existence and that the capacities represented in these inventions and discoveries did not occur among the people of a single culture, or race, nor in an integrated series of learning situations. Whatever may differentiate human learning abilities from those of other animals, such abilities must be assumed to be part of our human nature as are man's capacities, under the necessary motivational conditions, for transcending what he has learned through the reorganization of his knowledge and experience.[187] This would hardly be possible were

[185] Ernest R. Hilgard, *Theories of Learning,* New York: Appleton-Century, 1948, pp. 329–330.

[186] Harry F. Harlow, "Learning Theories" in *Current Trends in Psychological Theory,* 1951, p. 82.

[187] Although H. G. Barnett (*Innovation: The Basis of Cultural Change,* New York: McGraw-Hill, 1953) does not explicitly phrase the question with reference to human nature, nevertheless, the fundamental position he assumes and the fact that he has attempted to formulate "a general theory of the nature of innovation" strikes a new note in anthropological inquiry which is also bound to be of interest to sociologists and psychologists. "Students of the history of cultural growth," he says, "have been forced to answer the question of human inventiveness in their efforts to interpret the meaning of multiple occurrences of the same phenomena in different ethnic groups" (p. 18). But back of all interpretations (not only diffusionist or evolutionary, but the more intermediary position of American ethnologists) there lies the tendency to accept "one of two antithetical propositions relative to human inventiveness" (p. 17): human beings are believed to be either fundamentally creative or uncreative. One of the major difficulties that has stood in the way of resolving this antithesis has been conceptual. An understanding of the sources of cultural change at the grass roots level has been blocked by definitions of invention that "place the emphasis, not upon change itself, but upon estimates of its significance. They rate individual variation in terms of its degree, in terms of attitudes toward it, or in terms of its cultural consequences . . . a vast amount of such variation is ignored

it not for a unique but ubiquitous feature of man's psychological structuralization. This is the emergence of a dominant integrative center of the human personality and the development of ego-centered processes which permit self-objectification and likewise a basic orientation in a self-other dimension. A world of objects other than self, as well as the self as an object, necessarily emerge in the socialization process, although they have a varied cultural content given them.[188] Another unique facet of the psychological structure of man that distinguishes him from any other animal is the superego which, as Parsons says, "must be understood in terms of the relation between personality and the total common culture, by virtue of which a stable system of social interaction on the human level becomes possible." Ego-centered processes are modified by the internalization of attitudes and values derived from the cultural system in which the individual is socialized. The distinctive psychological structuralization of men, the novel qualitative features that characterize the psychodynamics of a human level of adjustment, combined with the generic capabilities of man that are rooted in the phylogenetic status of our species, lie at the very core of our human nature.

Viewed functionally and historically, it appears to be the indeterminate aspect of man's nature that makes him unique, the inherent potentialities of which, under the necessary motivational conditions, may lead to new and varied forms of social, cultural, and psychological adjustment. These potentialities can only be fully appreciated if, instead of looking backward from our present point of temporal vantage and contemporary knowledge, we select an earlier

---

because it does not fall within the arbitrarily established range of novelty. It is passed over if it does not elicit comment, or if it is not repeated or imitated, or if it otherwise is without lasting effects. The problem of change is thus shunted into the problem of measuring it." Barnett then goes on to point out that in actual fact "nothing is more common or certain than individual variability in concept or reaction." This is the source of variations that are labelled innovations because they acquire social significance. "Their fate is something apart from their origins . . . change as such is a universal phenomenon. Human beings have an infinite capacity for responding divergently. . . ." (p. 19.) If this proposition is acceptable, its significance for a science of social man cannot be overrated.

[188] Asch, *op. cit.*, Chapter 10, "The Ego," and A. Irving Hallowell, "Personality Structure and the Evolution of Man," *American Anthropologist*, Vol. 52, 1950, pp. 159–173; and "The Self and Its Behavioral Environment" in *Explorations*, Feb., 1954.

point in time and, looking forward imaginatively, compare what we know now about the culture history of man with what could only have been speculation then. Human behavior *is* relative to traditional culture patterns and historic circumstances but, at the same time, it is relative to unrealized potentialities that inhere in man's human nature and which permit the emergence of novelty in his mode of life. One of the great weaknesses of theories of cultural evolution which have stressed a series of inevitable stages in linear succession, as well as of diffusionist theories and doctrines of geographic or cultural determinism, has been an inadequate conception of the distinctive psychological nature of man.

Perhaps it is characteristic of man to be always different yet always the same. Perhaps this is what anthropologists have sensed without formulating it, since in moving from one people to another the field worker always has assumed that there were both psychological and cultural constants to be expected: identifiable emotions such as sorrow or hate, self-awareness and reflective thought, a scheme of moral values, a world view, tools, etc. Perhaps Pascal went to the heart of the matter when he said that the "nature of man is his whole nature." Eighteenth-century philosophers looked for what was common to man everywhere and identified what they found with the "natural" in contrast with customs that were local and provincial and consequently not related to human nature. On the other hand, anthropologists were, at first, more concerned with cultural differences than with cultural universals and they have been inclined to overstress cultural relativism. But cultural diversities and common denominators of culture[189] are part of the total human picture; both categories of phenomena must be related to the whole nature of man. Experimental psychologists once sought mainly for constancies, for responses that could be directly related to man's innate biological equipment and which could be taken as representative of the functioning of the human mind under the controlled conditions observed. They ignored the study of man as a social being and the varieties in his culturally constituted mode of life. But can human nature be exclusively identified with what is biologically innate and invariant, when man, the culture-building animal, is constantly reinterpreting his experience, reconstituting his world view, inventing

[189] George P. Murdock, "The Common Denominators of Cultures" in R. Linton, ed., *The Science of Man in the World Crisis*, New York: Columbia University Press, 1945.

new technologies, and adjusting his behavior accordingly? Philosophers once stressed the rational capacity of man; psychoanalysts have demonstrated the depth of man's irrationality. But cannot potentialities for rational behavior be a part of man's nature at the same time that unconscious impulses may limit his rationality? Neither rationality nor irrationality taken alone expresses the whole nature of man. Long before the days of psychoanalysis Nietzsche seemed to have sensed the issue: "Man in contrast to the animals," he said, "has developed a profusion of contradictory impulses and qualities. Thanks to this synthesis within his character he could become the master of the world!" A scientific approach to the question of human nature demands further knowledge of all the attributes that underlie the manifold adjustments of a creature that exhibits so complex a mode of living as man.

CHAPTER 7

Sociology and Psychology

THEODORE M. NEWCOMB

I. Elements and Systems

IF EACH OF THE several disciplines of human behavior took as its microcosm its neighbor's macrocosm, and its alternate neighbor's microcosm as its own macrocosm, their problems of articulation would be relatively minor ones. Such a well-structured community of disciplines might be diagrammed, for present purposes, somewhat as in Figure 1. According to this arrangement, Discipline B (perhaps referred to as social psychology) is primarily concerned with roles as systems constructed of single behaviors as elements. For Discipline A, however (which might be labeled individual psychology), a single behavior itself constitutes a system whose component elements are intra-individual events, either psychological or neurological. Similarly, the system of Discipline B becomes the element of Discipline C, whose specialized problem is that of investigating the manner of interdependence of roles in role systems. Each discipline, in short, would be taking a neighbor's system as its own element and/or another neighbor's element as its own system.

FIGURE 1

Elements:	intra-personal processes	behaviors	roles	role systems
Systems:	a behavior	a role	a role system	a societal system
Discipline:	**A**	**B**	**C**	**D**

As things have in fact turned out, the division of disciplinary labor has followed no such neatly patterned scheme. As the labels have customarily been applied, each field has tended to encroach upon one or both of its neighboring fields. The boundaries represent uncertain combinations of calculated surveyors' lines and ancient cowpaths legitimized by custom; many claims, moreover, are contested. There are advantages, nevertheless, in regarding an eventual science of man as one whose subsciences could be thus rationally related to each other. My present concerns are not so much those of avoiding jurisdictional squabbles as those of examining the conceptual relationships among neighboring subsciences.

Even if one prefers some other model to that described in Figure 1, there are relationships in common at every pair of adjacent levels of inclusiveness. At each level the specialist is less concerned with the "internal" structure and functioning of his elements than with their interrelations; for certain purposes, at least, he accepts them as undifferentiated entities. By the same token, he has only a limited interest in the relatedness of what is for him a total system to other like systems. At the same time, he tends to be critical of his neighbors in characteristic ways: of the more inclusive specialist, for casualness in taking the properties of his elements for granted, and failing to realize that properties vary with internal structure and process; of the less inclusive specialist, for ignoring influences "from above," according to which intra-system relationships are in part controlled by inter-system demands. At any level the specialist is probably aware that both kinds of criticisms apply to himself; he concedes the importance of "being informed" at adjacent levels, but reminds himself and his critics that one cannot work intensively at every level simultaneously.

FIGURE 2

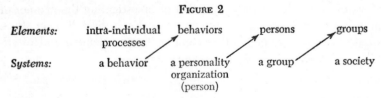

Elements:	intra-individual processes	behaviors	persons	groups
Systems:	a behavior	a personality organization (person)	a group	a society

There is much to be gained by understanding such inter-level influences, but the obstacles along the path toward a unified science of man are not of this simple order. Part of the difficulty, from the point of view of this chapter, is that sociologists and psychologists have different blueprints for the ideal division of labor. Figure 1, which might emerge from a sociologists' planning fest, may be contrasted with the characteristically psychological approach as outlined in Figure 2. If our disciplines had followed this particular pattern of differentiation, they might have been labeled, respectively, psychology of behavior, psychology of personality, social psychology, and sociology. This series of increasingly inclusive levels is based upon the common principle of behaving entities. In Figure 1, on the other hand, it is the identity of the behavior, not of the behaver, which is the concern at each level of inclusiveness. As I shall try to show, this fundamental difference in approach has done much to keep psychologists and sociologists from borrowing from each other.

Associated with this difference in relative emphasis upon behaver and behavior are two other differences. Psychologists working at any of the first three levels in Figure 2, tend to be alerted rather toward the left (i.e., toward less inclusive levels) than toward the right; and sociologists, who may be specializing at any except the first level in Figure 1, tend to be more sensitive to influences from the right. This, of course, is only a rough generalization. Insofar as it is a justifiable one, the distinctive trends represent tendencies to take as frame of reference, at each level, the most distinctive portion of the total field —distinctive not only in the sense of being farthest from the other field, but also in being most fully developed, and earliest developed. Crudely speaking, psychology got started from the left, sociology from the right.

A second characteristic difference has to do with relative emphasis upon historicity. Psychologists, being concerned with behavior as a function of the behaving entity, face the problem of continuity over time of that entity. It is almost a hallmark of their legitimacy that they are forever returning to their central problem—learning. Sociologists, who tend to view individual behavior as a function of the situation (the behaving entities being, within limits, interchangeable), are thus led to concentrate upon properties of existing situations. This is not to say, of course, that sociologists are indifferent to problems of individual continuity. When they turn up such problems, however, they often become quasi-psychological (e.g., "the socialization of the individual") or quasi-anthropological (e.g., culture as that which is socially transmitted). Nor is there any intended implication that ahistorical approaches are foreign to psychologists; the work of Kurt Lewin is eloquent testimony to the contrary. It is no accident, however, that Lewin (and other neo-Gestalt psychologists, though less conspicuously) found the ahistorical approach readily applicable to the study of individuals as group members, and thus went on to the study of groups themselves.

Between sociology and psychology, then, there are genuine divisive forces. Like two formerly insulated societies pushed by population growth, they have spread over into the same intervening territory, and both are suffering the pains as well as the gains of culture contact. Those who would maximize the gains must familiarize themselves with the culture and recent history of both societies. A brief overview of the recent "interpenetration" of sociology and psychology therefore follows, particularly from the point of view of sociological contributions to psychology.

II. Extensions of Microsociology and Social Psychology toward One Another

Traditionally sociologists have not limited their concerns to any single level of analysis, their jurisdictional claims being virtually coterminous with "the proper study of mankind"—or at least the study of man in all his social relationships. Our library shelves probably contain quite as many sociological treatises on social determinants of the individual personality as on the structure of society. If it had proved possible for an earlier generation of sociologists to develop a body of theory which was both reasonably complete and internally self-consistent, sociology itself would have been on the high road to a unified science of man.

A very large proportion of sociological activities may be thought of as centering about one or the other of two nodes. One of these is the interpersonal interaction process, in which the individual, as source and as object of social stimulation, is the central focus of concern. The other node is that of societal structure, and its focal problem is the interrelationship of differentiated collectivities. The former, which is loosely referred to as "social psychology" (or sociological social psychology), I shall label *microsociology*, since it deals with the smallest unit accessible to sociologists, the single person. *Macrosociology*, commonly referred to as the field of social organization, or ecology in its most inclusive sense, deals with collectivities as units. The properties of collectivities vary, of course, with the nature of interaction internal to them; interaction processes, in turn, are made up of interrelationships among individual psychological processes. Thus traffic between macrosociology and psychology is routed through microsociology, and it is the latter, more intimate connection which is the primary focus of the present analysis.

Psychologists, till recently at least, have tended to work at lower levels of inclusiveness than have sociologists. For the most part, when they have dealt with groups, they have clung to the model of the individual as a stimulating, stimulated, and responding organism, two or more of which together constitute what was for them a quasi-fictional group. A whole social psychological epoch is epitomized by the title of the dissertation of one of F. H. Allport's students, R. L. Schanck: "A study of a Community and Its Groups and Institutions Conceived of as Behaviors of Individuals."[1] To most psychologists, including social psychologists, the phenomena of social behavior did not seem to require a new model, nor even many accessories to the existing one. Social stimuli, and responses thereto, are phenotypically distinguishable from others, to be sure—but this is true of all classes of stimuli, and hence no special theoretical challenge was felt. It is thus not surprising that the place accorded to social psychology within the psychology family was quite different from that granted to the corresponding child of the same name in the sociology family. Whereas sociologists tend to regard social psychology as a major subdivision within their theoretical field, psychologists have usually viewed it as an applied field, like industrial or educational psychology; thus theoretical traffic between psychology proper and social psychology was almost entirely from the latter to the former.

Nevertheless, psychological social psychology (to which I shall henceforth refer simply as "social psychology") has begun to develop an autonomy of sorts, and with it theoretical ambitions of its own, which have begun, at least, a reverse direction of traffic toward general psychology.[2] More psychologists than ever before are now interested in one or another aspect of the social determinants of behavior. It is to those who are interested in such matters that the term "psychologist" is henceforth applied, whether or not they label themselves "social psychologists."

[1] *Psychological Monographs*, 1932, 43, No. 195.

[2] I have tried to point to some of the theoretical contributions of social psychology to one of psychology's major problems, motivation, in *Current Theory and Research in Motivation, a Symposium*, University of Nebraska Press, 1953.

Psychologists have commonly taken a dim view of the contributions of microsociology. They have pointed out that the study of human interaction is necessarily a psychological study, in part at least, whether it is so labeled or not, and that much of the psychologizing which has gone into microsociology has been distinctly amateurish and casual. The less offensive forms of this particular sin are those of sheer naïveté and simple unawareness that psychological assumptions are being made. Less forgivable are persistence in psychological error and outright dogmatism; favorite examples are the minimizing of biological and genetic determinants of behavior, and the assumption of passivity on the part of organisms which merely "absorb" folkways or culture, leading to a view of personality as merely "the individual aspect of culture." Psychologists have complained, too, that microsociological contributions are suspect because so often proffered without benefit of experimental evidence, or even without systematic and controlled evidence of any kind.

Such allegations, regardless of their correctness, are symptoms of restricted communication. The more sophisticated microsociologists have least often laid themselves open to such charges, and the psychologists who have taken the trouble to read widely and representatively have been least ready to level a general indictment. Unfortunately, however, misunderstanding here as elsewhere has too often led to withdrawal and further misunderstanding; the allegations have served to aggravate the state of malcommunication out of which they grew. Such separation has been characterized by the co-existence of a *psychological* social psychology and a *sociological* social psychology (microsociology) which, at worst, had almost nothing in common.

This state of affairs, however, was more characteristic of the years between the two world wars than it has been of the recent past. This is not the place to note the historical forces which have led to many signs of rapprochement. Microsociological treatises have become more psychologically sophisticated and at the same time more solidly grounded on empirical data. Psychological studies, meanwhile, have become increasingly oriented toward the interactional context in which so much of individual behavior occurs. In short, students of human behavior who call themselves psychologists (usually *social* psychologists, to be sure) have discovered that there is something to be learned from sociology. It is with this development that the present chapter primarily deals. Insofar as this development can be understood, it will go far toward showing why the lines between psychological social psychology and microsociology are tending to disappear.

Sociology has already made certain contributions to psychology, in this sense; I shall briefly mention some of them, before going on to mention some other potential contributions. The first and most important of the contributions already made, in the sense that it has had considerable effect upon the thinking of many psychologists, is a very general and inclusive one; in a way it includes those which I shall mention later. It is nothing less than the necessity of coming to terms with the ineluctable fact of groups. Few psychologists of a generation or two ago would have cared to deny, I suppose, that it is possible to distinguish among collectivities of persons. Nevertheless they felt themselves helpless to handle the phenomenon. Their dilemma was that on the one hand they could not apply strictly psychological concepts to a group (after all, as F. H. Allport once observed, a group does not have a nervous system), and on the other hand they were unwilling to make loose and reckless extrapolations of their individual concepts (like "group mind").

The path of escape from this dilemma, as so often in the history of science, was to take a fresh look at the facts. Several sets of facts available to psychologists pointed to the possibility of viewing an individual as a unit in a more inclusive entity. Those who were aware of developments in biological theory (who had read C. M. Child, for instance, or G. E. Coghill) knew that identical living cells would develop differently if transplanted into different parts of the same organism. Those who had dispassionately examined some of the early experiments of the Gestalters (perhaps Wertheimer or Köhler) became familiar with problems of part-whole relationships. There were many analogies from physics and chemistry. There was nothing inherently contra-scientific, in short, about the possibility that the behavior of persons varied with the organization-system of which those persons were parts. No single sociologist had more influence in this direction, probably, than C. H. Cooley whose calm and nondefensive insistence on the indivisibility of "Human Life" was belatedly discovered by many a psychologist. Almost simultaneously with their discovery of Cooley, many of them found Kurt Lewin driving home a similar point, though in a very different language: groups are "real" if, as, and when they have "real" effects. His ingenuity in demonstrating the "reality" of such effects by experimental means has done much to establish the point among psychologists.

If this contribution had gone no further, it would have amounted to little more than a point of view—and one, moreover, which psychologists interested in social behavior would necessarily have come to by themselves, sooner or later. The necessity of taking this point of view carried with it the necessity of asking more questions. The "real effects" of groups upon their members cannot be accounted for in the traditional psychological way of regarding each member, one at a time, as an organism responding to a sequence of social stimuli, since in this way the group itself never appears, except synthetically. Such considerations led to inquiries about the relationships among the elements (persons) of which the system (group) is made up. As Cassirer long ago pointed out,[3] the history of the older sciences has been characterized by progressing from problems of constancy of elements to those of constancy of relations.

Here microsociology had something to offer. Particularly at the hands of G. H. Mead, the notion of response gave way to that of interaction. There are constant relationships between behaver and observer—not only because of personal continuity through time, as the same person is now observer and an instant later observed, but even simultaneously as the same person observes his own behavior and in so doing identifies himself with others who observe him. The constant relationship is one of participation in a common matrix ("the generalized other"), its constancy maintained through communication by means of symbols which point to common anticipations on the part of the interacting persons.

Curious as it may seem at first, it was the microsociologists rather than the psychologists who first noted the crucial significance of speech in social behavior. True, there is a psychological tradition of interest in the problem of speech acquisition, but with the notable exception of Piaget, psychologists have in general failed to extend this interest to the problem of socialization. Psychologists, continuing to see the problem of the constancy of relationships as an intra-individual one, tended to see speech as one among many forms of adaptive behavior. Microsociologists, seeing their problem as an inter-individual one, viewed speech as a communicative device, analogous to the transmission of nerve impulses for the physiologist,

[3] E. Cassirer, *Substance and Function*, translated by W. C. and M. C. Swabey, Chicago: Open Court, 1923.

and found in it an essential mechanism—for some of them, *the* essential mechanism[4]—for maintaining the integrity of the multi-person system. Small wonder, then, that the contribution here has been chiefly from microsociologists to social psychologists and not vice versa.

A closely related contribution—the concept of "the social self"—is one which might more plausibly have been expected to come from psychologists. After all, the phrase was originally William James's. J. M. Baldwin and John Dewey, psychologists of some eminence, made much of the concept, each in his own way. But the historical fact is that their contributions have penetrated the sociological tradition far more deeply than the psychological. Cooley, at about the same time as Baldwin and Dewey, was saying, in effect, that even the most personal and private aspect of a person, his very self, was a social product. And Mead, shortly thereafter, was finding in the self the microcosmic counterpart of society. Here, as in the case of speech, the difference between psychologist and microsociologist lay in the fact that the latter was compelled to deal with the fact of interaction, as the former was not. The concept of the self provided an indispensable intervening variable to account for the constancy of interpersonal relationships—intervening in the sense that the self is both product and determiner of those constancies.

But the fact of collective constancies, as well as that of individual constancies, must be faced; after all, properties of groups often persist as their members change. From the point of view of the power of a group's constancies over its individual members, this is a microsociological problem, though the authors of concepts like collective representations and folkways and mores were not primarily interested in problems of interpersonal interaction. More directly influential upon psychologists, probably, was the conceptual framework of Thomas and Znaniecki, the usefulness of whose interdependent concepts of social value and individual attitude was widely noted. For reasons noted below, and particularly as a result of the writings of M. Sherif,[5] many psychologists have preferred to handle the fact of such group constancies which have power over their members in terms of the concept "social norm."

[4] Cf. A. R. Lindesmith and A. L. Strauss, *Social Psychology*, New York: Dryden Press, 1949.

[5] Especially *The Psychology of Social Norms*, New York: Harper and Brothers, 1936.

One other contribution—the concepts of position, or status, and role—is of such central importance to the psychologist concerned with social behavior that a special section is devoted to it below.

At their own level, then, and with remarkably little help from psychology, microsociologists—of the 1920's, say—had developed a framework for handling many of the observed facts, conditions, and consequences of interaction among persons. For them the existence of groups as entities in their own right constituted no problem; it was an initial assumption. Groups were characterized by their own properties, including the power to influence or "control" their members. Individuals, too, had their constant properties, which were chiefly outcomes of participation in group life. The process by which both kinds of constancies were maintained was that of social interaction, in which language played an indispensable part.

The fact that such a framework existed, however, was no guarantee of its use by psychologists, as their interests turned in social directions; a potential contribution is not an actual contribution until it is exploited. During the 1920's—and indeed, with minor exceptions until the late 1940's—psychologists made little use of the potential contribution. Most of them were simply not interested in social psychology; until the late 1930's there had been no important textbooks by psychologists except those by McDougall and by F. H. Allport. Social psychology was taught in few departments of psychology; not until 1947 was "social psychology" separately categorized in the *Psychological Abstracts* (although a section on "social processes" had long been included). The few psychologists who were interested in the area, moreover, were approaching it from their own direction. This was hardly surprising, of course, but it is nevertheless instructive to inquire why the potential contribution which was earlier almost ignored, later became an actual contribution.

The earlier indifference to the contribution was paralleled by the difficulty in recognizing anything psychological about it—or anything acceptably psychological, at any rate. Naïve operations of transplanting individual psychological concepts into phenomena of group behavior (as LeBon seemed to have done) were understandably repugnant to psychologists. Those few of them who had read Durkheim could not be quite sure whether he was committing the same fallacy or not, and there was little in a post-Wundtian world to attract them to him in any case. Sumner dealt with individuals only as psychologically undifferentiated pawns in the grip of custom.

Even the microsociologists who attempted to work more systematically at the level of interaction among individuals, and who were therefore somewhat closer to psychological problems, seemed remote in their approaches. The study of suggestion, imitation, and sympathy ("the classic triad") seemed legitimate enough to the psychological student of social behavior, for such processes could be investigated in terms of individual stimulation and individual response, and had been so conceived by McDougall and F. H. Allport. Tarde's "laws of imitation," however, had nothing to say about individual psychological variables, and Ross's "explanations" of "planes and currents" seemed patently *ex post facto* descriptions of events selected to illustrate certain classes of phenomena, rather than a statement of conditions under which persons influence one another in specified ways. Cooley came more closely to grips with the empirical facts of face-to-face interaction but, quite apart from the fact that he seemed to be avoiding psychological terminology, his approach was so global that it could neither be denied nor put to very practical psychological use. Mead's "social behaviorism" sounded attractive in a post-Watsonian era, but his writings often gave the impression that individuals were to him interchangeable specimens of gutless creatures, devoid of sensation, perception, motivation, and emotion and innocent of susceptibility to conflict. Many psychologists have noted that comparatively little empirical research has stemmed from the Meadian system of thought.

Such contributions eventually made their impact, but not until psychologists had built their own bridges out to meet them. These bridges made possible a kind of rapprochement, or rather, perhaps, a reciprocal acculturation. Important among them were extensions of psychological theory itself, the consequences of which for social psychological theory were unpremeditated but nonetheless significant. The substantial and interdependent developments in theories of motivation, perception, and learning during the past quarter-century cannot even be outlined here, but a few of them will be hinted at as some of the bridge-building toward microsociology is described.

An important and relatively early development was the relationship between social norms and perceptual processes. In particular, the work of two widely read psychologists, Piaget and Sherif, forced the attention of many of their colleagues upon the fact of social norms. Piaget's patient documentation of the phenomena of social perspectivism in children could not be ignored.[6] Sherif's experimental demonstration that the communication of judgments of an ambiguous event resulted in a common judgmental norm, individually maintained by those who had communicated about it while at the same time an obvious group product, had enormous influence. To many microsociologists the experiment seemed an elaboration of the obvious, but to countless psychologists it marked a turning point, showing that the effects of social interaction could be so basic as to modify the fundamental, psychological process of perception itself. Psychologists, like their microsociological brethren, did not need to be told *that* group norms had power over group members, but they did need to be shown *how* the process operated, and preferably by experimental demonstration. By grounding the fact in a familiar psychological process, this single experiment went far toward "legitimizing" group norms and making them theoretically accessible to psychologists. And insofar as this happened, many of the related contributions of microsociologists became accessible to them.

Psychological extension of the self concept represented another important bit of bridge construction, work on which had been virtually abandoned (as noted by G. W. Allport[7]) for some decades. It is no accident that both Piaget and Sherif, finding problems of perspectivism and perception inseparable from those of self-cognition, took up the latter as a central problem. It was one of Piaget's most solid contributions, buttressed by observations obtained in his early studies of language, judgment, reasoning, etc., that self-other differentiation develops step by step with the decline of solipsism (egocentricity) and the substitution therefore of social rules based upon the understanding of reciprocity. Ego waxes as egocentrism wanes. This approach provided direct entry to the vestibule of the microsociologists, and it is of considerable interest that in two early post-war volumes[8] Sherif has whole chapters that read almost like microsociological treatises.

[6] See especially the concise summary of his theoretical position in *Factors Determining Human Behavior*, Harvard Tercentenary Publications, Cambridge: Harvard University Press, 1937.

[7] "The Ego in Contemporary Psychology," *Psychological Review*, 1943, 50, 451–478.

[8] *An Outline of Social Psychology*, New York: Harper and Brothers, 1948; and (with H. Cantril), *The Psychology of Ego-Involvements*, New York: Wiley, 1947.

A wide range of influences, including his own writings, contributed to the re-emergence of the self which Allport noted in 1943. Freudian, and especially "neo-Freudian" influences had a considerable impact. Among the latter the work of H. S. Sullivan probably seemed most sophisticated to both psychologists and sociologists.[9] Many psychological students of personality were influenced, too, by the "researchability" of self-percepts, as most notably illustrated by the research of C. R. Rogers and his associates.[10] As of today, at any rate, few students of personality as "the individual's organization of predispositions to behavior"[11] could ignore the central, organizing functions of the self.

Many psychologists would be inclined to judge the relevance of a given problem to psychological theory in terms of the applicability of learning theory to it. At least two applications and one extension of learning theory have brought it closer to microsociology during the quarter-century. As early as 1924 F. H. Allport noted, in his *Social Psychology*, that many of the phenomena of suggestion and imitation could be regarded as conditioned responses; seven years later G. and L. B. Murphy were distinguishing (in *Experimental Social Psychology*) several variants of these and other forms of social interaction in terms of differing learning conditions. In 1941 Miller and Dollard, in their *Social Learning and Imitation*, made a somewhat comparable distinction among the varieties of imitation, supporting their "four factor learning theory" by experimental demonstrations of imitation in rats and in children. A second, more highly generalized application of learning theory to problems of interaction is to be seen in R. R. Sears's[12] analysis of "the dyadic unit"—i.e., "one that describes the combined actions of two or more persons." The concept of "expectancy" (of which more below), derived from the "monadic" principles of learning, is crucial to this account.

Learning theory during the quarter-century has become more closely related to the theory of motivation, as well as to that of per-

[9] Note the glowing tributes by G. Murphy, by Otto Klineberg, by L. S. Cottrell and N. N. Foote, and by O. S. Johnson in *The Contributions of H. S. Sullivan*, edited by P. Mullahy, New York: Hermitage House, 1952.

[10] Cf. his *Studies in Client-Centered Psychotherapy*, Psychological Service Center Press, 1952.

[11] My own preferred definition; cf. *Social Psychology*, New York: Dryden Press, 1950.

[12] His presidential address before the American Psychological Association, 1951; *The American Psychologist*, 1951, 6, 476–482. It is of interest to note, in passing, that E. R. Hilgard's presidential address, in 1949, was devoted to the topic of the self.

ception, and an important extension has had to do with the learning of motives. Hull and his associates at Yale have had much to say on this problem which, incidentally, is prominently treated in the Miller-Dollard volume on imitation. Not all psychologists would be satisfied with their theory that "drive value is acquired by attaching to weak cues responses producing strong stimuli," but all of them have had to wrestle in one way or another with motive acquisition as a problem in learning theory. As learning theory becomes more capable of handling this problem, it obviously becomes more possible for social psychologists to make use of sociological contributions —e.g., by studying the properties of a structured social environment as conditions of motive acquisition.

Finally, as a theory-building bridge of very general character, the development of "field theory" takes an important place. According to Lewin, whose influence upon post-war social psychologists was very considerable, "field theory is probably best characterized as a method; namely, a method of analyzing causal relations and of building scientific constructs."[13] Lewin's insistence upon "the characterization of events and objects by their interdependence rather than by their similarity or dissimilarity of appearance" was influential because his derivations and hypothetical predictions so often "paid off" in the coin of theory and experiment. Lewin, more than any other single figure of his time, helped his fellow psychologists to see the variety of ways in which the properties of single persons varied with the manner of their interdependence with other persons in groups, and in so doing he made it both more possible and more necessary for psychologists to turn toward sociologists for some of the facts of life at the group level.

III. Social Structural Determinants of Individual Properties

Such were some of the influences upon psychologists which brought them within shouting distance of sociologists. Closer proximity did not necessarily mean that communication was easy. Not only were there new concepts and strange vocabularies to be learned; there was often interference from old habits, as familiar phenomena were viewed in a different perspective. Nevertheless, those psy-

[13] *Field Theory in Social Science,* edited by D. Cartwright, New York: Harper and Bros., 1951.

chologists who ventured out on the extending bridges soon found readily exploitable bodies of data and theory.

There is no harder lesson for the psychologist to learn, probably, than that of viewing persons as functionaries in a group structure rather than as psychological organisms—i.e., as parts rather than as wholes, and as parts which, within limits, are interchangeable. The fact that the parts do not "stay put" within the same whole, as do the parts of single persons, makes the lesson the more difficult for the person-oriented psychologist. Once this lesson is learned, however, the facts of social structure become available; a social system is seen as made up of differentiated parts, the orderly relationships among which, rather than the personal identity of which, become the major object of concern. The possible contributions of a micro-structural approach are obvious. At a simple, phenotypic level the properties of persons can be investigated as varying with their positions in family or factory structures. At a more genotypic level properties of persons can be seen as functions of abstractable dimensions of social structure, such as authority or prestige-status, which are diversely associated with such phenotypes as sex, age, and occupation.

No richer source of data on this problem has yet appeared than *The American Soldier*, on many a page of either volume of which one can find evidence of how attitudes vary with rank in the army structure. H. Speier[14] has assembled some dozens of comparisons of attitudes of officers and enlisted men which go beyond the mere demonstration of differences, and eloquently support the hypothesis that "the difference between opinions on the same subject matter expressed by groups high or low in power, privilege, or prestige will increase as the subject matter is more closely and directly related to the status characteristics and relations of the group." (E.g., differences are much greater concerning attitudes toward special privileges for officers than concerning the difficulties of defeating the enemy.) This fact, together with that of systematic attitudinal "gradients" in ascending the scale of rank, makes it abundantly clear that such attitudes are not only properties of single persons but also properties of parts of systems.

The social psychologist is bound to be interested in such findings,

[14] In *Continuities in Social Research: Studies in the Scope and Method of "The American Soldier,"* R. K. Merton and P. F. Lazarsfeld, eds., Glencoe: Free Press, 1950.

but if he has taken his social-structural lesson to heart he will not be content with concluding that individuals have "interiorized" the social norms of their peer-groups. He will go on to the macrostructural fact of relations among rank-differentiated peer-groups. Somehow the single person, as an elementary part, is being influenced by structural relations of parts larger than himself. The group norms themselves—almost as if disembodied—are adaptations to intergroup relatedness. To the psychologist, of course, they are not outcomes of disembodied processes, but his account of the very-much-embodied processes must be embedded in a macrostructural context. As suggested below, such an account must (in the writer's judgment) be given in terms of such interactional concepts as role, communication, and consensus.

The phenomena of individual differences associated with social classes have, more than any others, probably, forced social psychologists to face such problems. The work of A. Davis and his associates has been particularly influential. Class-associated individual differences—whether of vocabulary, child-rearing practices, personality organization, or attitudes toward national affairs—can be "explained" only at the most superficial level if *inter*-class relationships are ignored. The reciprocity of inter-class roles, their social reinforcement by members of both own and other class groups, barriers to inter-class communication, "socially shared autisms" (to borrow G. Murphy's phrase) with reference to adjacent classes—these and other phenomena of inter-class relationship are essential ingredients in any adequate account of the forces whose end results are class-associated individual properties.

The concept of "reference group" has proven a useful one to many social psychologists, though its full sociological implications have yet to be empirically investigated. As is proper with intervening variables, its first systematic use as an explanatory concept was mothered by necessity, when the authors of *The American Soldier*[15] found it necessary to call upon a hypothetical "relative deprivation" to account for several sets of findings. Specifically, they hypothesized that satisfactions and dissatisfactions in a variety of contexts varied with relevant properties attributed to groups with whom the individual compared himself. More recently H. H. Kelley[16] has dis-

[15] Who did not, however, use the phrase, which was first employed by H. Hyman in "The Psychology of Status," *Archives of Psychology*, 1942, No. 269.

[16] In *Readings in Social Psychology*, Revised edition, edited by G. E. Swanson, T. M. Newcomb, and E. L. Hartley, New York: Henry Holt, 1952.

tinguished between this *comparative* function of reference groups and a *normative* function, referring to the influence of norms attributed to a group—an influence whose effects are not limited to self-judgments.

The most systematic attempt to develop a theory of reference group behavior is that of Merton and Kitt[17] (who limit themselves, however, to the former sense of the term). Perhaps their most significant contribution, for present purposes, has to do with the problem of multiple reference groups—especially "conflicting" ones. Certain findings concerning the morale of overseas noncombatants could best be explained by the hypothesis that, relative to soldiers still at home, overseas noncombatants were considerably disadvantaged, but not relative to overseas combatants. But, if so, why were these particular comparisons made? Merton and Kitt's answer, briefly, is that the reference groups are socially structured ones. "The reference groups here hypothesized . . . are not mere artifacts of the authors' arbitrary scheme of classification. Instead, they appear to be frames of reference held in common by a proportion of individuals within a social category sufficiently large to give rise to definitions of the situation characteristic of that category. And these frames of reference are common because they are patterned by the social structure." They are "used" psychologically, in short, because they have real effects upon those who use them. Thus the same Lewinian logic which compels the social psychologist to accept groups as "real" insofar as they have "real" effects also compels him to take social structure as equally "real."

Many psychologists who have been confronted by the facts of social structure have been deflected from further investigation of those facts by the psychological truism that, within certain limits, the effects of any influence vary with the manner in which that influence is experienced. As between class strata, for example, individual attitudes and behaviors with respect to members of other class strata are determined not so much by the objective facts of macrostructure as by the perceived properties of members of other strata. And these perceived properties are often determined far more by interaction within the individual's own stratum than by familiarity with "the facts." With regard to attitudes of American whites toward Negroes, this position is clearly stated in the concluding remarks of an early

[17] In Merton and Lazarsfeld, *supra*.

and influential study by a psychologist:[18] "It seems that attitudes toward Negroes are now chiefly determined not by contact with Negroes but by contact with the prevalent attitude toward Negroes."

With such a conclusion there can be no quarrel *at a social psychological level;* the documentation of such a conclusion in 1936 was, in fact, a major contribution. But it leaves unanswered, of course, the question as to why "the prevalent attitude"—i.e., prevalent among whites—is what it is. Let us concede immediately the advantages, and even the necessity, of a division of labor. Neither the social psychologist nor any one else can investigate everything simultaneously; at any level of operation it is necessary to take certain phenomena as given. Nevertheless the fact remains that "the prevalent attitude" almost certainly represents an adaptation by one part of a macrostructure (whites) to another (Negroes). Thus the question is bound to arise: How do properties of individuals (attitudes of whites toward Negroes) vary with the processes by which the microstructure in which they as individuals have positions, adapt to other microstructures? If it turns out that there are significant covariations of this sort, then the social psychologist is losing something by not taking them into account.

An example of an attack upon this kind of problem is to be seen in Merton's "Social Structure and Anomie"[19] in which the problem posed is that of "how some social structures exert a definite pressure upon certain persons in the society to engage in nonconformist rather than conformist conduct." Starting with the central thesis that "aberrant behavior may be regarded sociologically as a symptom of dissociation between culturally prescribed aspirations and socially structured avenues for realizing these aspirations," he concludes that: "It is only when a system of cultural values extols, virtually above all else, certain *common* success goals *for the population at large* while the social structure rigorously restricts or completely closes access to approved modes of reaching these goals *for a considerable part of the same population,* that deviant behavior ensues on a large scale" (author's italics). The significance of this kind of theorizing, for present purposes, is its analysis of differentiation among parts of the macrostructure. Given a macrostructure with these kinds of part-whole relationships, and given human organisms to whom certain

[18] E. L. Horowitz, "The Development of Attitude toward the Negro," *Archives of Psychology,* 1936, No. 194.

[19] In his *Social Theory and Social Structure,* Glencoe: Free Press, 1949.

"standard" motivations are attributed, different properties of parts (e.g., differential crime rates among ethnic or class groups) are predictable.

In what sense are such macrosociological contributions of use to psychology? Primarily, I suspect, by way of suggesting hypotheses for study at psychological levels. If, for example, differences between structurally differentiated parts are maintained by the exercise of force, authority, or power exercised primarily by one of those parts, factors are introduced which both the social and the personality psychologist will sooner or later want to study. Although Marx had something to say about this problem, as applied to social classes in industrial society, he did not raise—much less answer—basic psychological questions as to the conditions under which force is perceived as such by members of different classes or, if so perceived, under what conditions varying kinds of individual behavior are most likely to ensue. G. C. Homans,[20] at the microsociological level, has raised this kind of problem and proposes various hypotheses concerning the effects of authority relationships upon the interdependence of "the elements of behavior"—activity, interaction, sentiment. E. Jaques,[21] in a volume which presents both a detailed case history and theory of the interpenetration of organizational form and individual motivation, attacks a similar problem in terms of facilitating or restraining "sanctions" applied by persons or (more commonly) by groups to persons, to roles, and to role systems carrying authority.

For these and for many other students of human behavior psychological problems have arisen in the course of studying social structures; hypothetical clues have been provided, moreover, for psychological investigation. The pursuit of such clues has only started, but two kinds of gains have already begun to appear. The first is a movement toward an eventual unity of science. As psychologists discover that they cannot answer their own problems without taking account of social-structural variables,[22] they begin to acquire a familiarity with sociological perspectives and vocabularies which

---

[20] *The Human Group,* New York: Harcourt, Brace, 1950.

[21] *Changing Culture of a Factory,* New York: Dryden Press, 1952.

[22] See, for example, D. Katz and R. L. Kahn in *Readings in Social Psychology,* G. E. Swanson, T. M. Newcomb, and E. L. Hartley, eds., New York: Dryden Press, 1952, who note "the inadequacy of applying the behavioral model of individual motivation to a social organization. . . . Recent industrial studies [they continue] argue strongly for the inclusion of social-organizational variables. . . ."

is a necessary though not a sufficient condition for progress toward a unified science.

Secondly, and more pertinently for the present chapter, the pursuit of sociological clues promises substantial gains *for psychology itself*. For example, as psychologists become aware of the inadequacies of "drive-centered" theories of motivation and learning,[23] they turn with new interest to formulations in terms of environmental "incentives." They may or may not turn to social-structural variables as determining many such incentives for humans, but those who do are most likely to be impressed by the need of theories of motivation and learning which are less exclusively drive-centered. Such theories will, of course, be "better" not just because they are "more social" but because they take account of a wider range of facts not adequately accounted for otherwise.

The psychological theory of expectancy offers another example. Among other reasons for a growing interest in this problem is the necessity for accounting in psychological terms for the observable facts of social interaction.[24] It is likely that psychologists would be forced to perfect a theory of expectancy even to account for rodent behavior, but it is also likely that the rich yield of empirical data provided by the observed forms of order and regularity in human social structures will contribute greatly to its development. It is no accident that much microsociological theorizing (by Parsons and Shils, most recently) has been carried on in the language of expectations.

In sum, those psychologists who have turned their attention to theoretical problems of social behavior have repeatedly discovered that order and regularity are to be found not only at the level of the organism but also at the level of collectivities. There are three possible answers to the problem of the interrelatedness of lawfulness at the different levels: it is pure chance; lawfulness at the more inclusive level is required by the nature of lawfulness at the less in-

[23] Cf. H. F. Harlow's "Mice, Monkeys, Men and Motives," *Psychological Review*, 1953, 60, 23–32, and D. O. Hebb's "Social Significance of Animal Studies" in *Handbook of Social Psychology*, G. Lindzey, ed., Addison-Wesley, 1953.

[24] R. R. Sears, for example, in the presidential address already cited, notes that "the factor responsible for maintaining the stability of the dyadic unit . . . is the expectancy of the environmental event." Sears takes as his model Hull's "anticipatory goal response," but other models (e.g., Tolman's) are equally applicable.

clusive level; or the reverse of this. The first of these is rejected on grounds of scientific faith; the second is hypothetically accepted but seems to be inadequate (e.g., alternate forms of social organization seem equally compatible with given forms of individual organization); the third alternative, that lawfulness at less inclusive levels is required by the nature of lawfulness at more inclusive levels, must therefore be explored also. It seems altogether likely that, with its exploration, there will appear new demonstrations of the manner in which whole-organizations limit and selectively influence various aspects of part-organization. With such developments, psychological theory itself will have been improved.

IV. Some Approaches to the Conceptual Analysis of Inter-level Influence

Such a development could hardly occur without some borrowing by psychologists of sociological concepts. Nor is it likely that the borrowing could occur without such elaboration of concepts as seems to make them more amenable to psychological use. The remainder of this chapter deals with the interdependent concepts of role, consensus, and communication, which are particularly appropriate for the rendering of social-structural facts into psychologically accessible data.

Psychologists—even those who use the prefix "social"—have been slow to adopt what would seem to be the obviously useful concept of role.[25] With rare exceptions, their organism-centeredness has apparently prevented their seeing its microstructural implications. The following definition[26] suggests the nature of the difficulty confronting psychologists: "A person's role is a pattern or type of social behavior which *seems situationally appropriate to him* in terms of the demands and expectations of those in his group" (my italics). Such an organism-centered definition is to be contrasted with one like the following, the locus of which is a collectivity: "We must de-

[25] Not until 1950, apparently, did any textbooks in social psychology by psychologists devote systematic treatment to the role concept: those by S. S. Sargent (Ronald Press) and by T. M. Newcomb (Dryden Press). M. Sherif (Harper and Bros., 1948) at about the same time, subsumed many role phenomena under the concept of ego-involvement. Even K. Young, originally trained in psychology, makes no systematic use of the role concept in his second edition (1944), except in the Meadian sense of taking the role of the other.

[26] By S. S. Sargent in *Social Psychology at the Crossroads*, J. H. Rohrer and M. Sherif, eds., New York: Harper and Bros., 1951, p. 360.

fine social role as an organized pattern of expectancies that relate to
. . . reciprocal relationships to be maintained by persons occupying
specific membership positions and fulfilling definable functions in
a group."[27]

The role concept is potentially attractive to psychologists primarily because it points to behavior on the part of individuals. It is
necessary to sociologists because it points to uniformities at one or
another collective level without regard to continuities of specified
organisms. If the psychologist is to profit from what the sociologist
has to say about microstructural determinants of behavior, he must
carefully distinguish between two kinds of problems. The first, by
which he is most apt to become intrigued, deals with social-structural
influences upon the individual organization of behavior; for this purpose the dependent variables are not roles, but *role behaviors*.[28]
(E.g., What are the effects upon personality organization of being
reared as an only son in a middle-class, urban, professional family
in contemporary America?) The second kind of problem has to do
with the nature of the processes by which interpersonal relatedness,
or organization, of behavior is maintained. For this purpose role
behaviors of two or more persons must be viewed as independent
variables at one moment and the next moment as dependent variables. For example, how can we account for stability and change in
the relationships between only sons of the kind described above,
and their mothers? To face such a problem forces us to regard mother
and son as somehow constituting a system, and to postulate some
kind of intervening variable which maintains the system, in the
sense of being a resultant of previous role behaviors of all the role-
takers involved, as well as determining their present role behaviors.
By a certain definition—and, as it seems to me, the necessary one—
it is the role which constitutes this intervening variable at the collective level, and, at the individual level, the role as perceived by the
role-taker.

Role, thus defined, is not behavior at all, but "a pattern of expectancies" regarding behavior—a multi-person pattern, the nature
of which determines today's role behaviors and is determined by

[27] E. L. and R. E. Hartley, *Fundamentals of Social Psychology*, New York:
Alfred A. Knopf, 1952, p. 486.
[28] I have tried to distinguish between these concepts in "Role Behaviors in
the Study of Individual Personality and of Groups," *Journal of Personality*,
1950, 18, 273–289.

yesterday's role behaviors. One of the major properties of such a pattern is that of the degree and content of inter-individual consensus regarding relevant role behaviors. Before turning to this role variable, however, a concrete example may be offered of how the sociology of roles has contributed to psychological thinking.

Psychological approaches to the study of leadership, until quite recently, were largely in terms of personal qualities—especially "dominance"—of persons holding designated kinds of positions.[29] Often with more references to the "pecking order" structure of barnyard fowls and of primates than to the social structure of human groups, such accounts were apt to include careful reports of frequency scores of various lists of leader traits. But the very diversity of the empirically reported traits of leaders made theorizing difficult. Interactional approaches, taking into account the reciprocal dependence of leaders and followers upon one another, became more common during the late 1940's. With varying degrees of sociological sophistication, social psychologists are now beginning to develop theories which explicitly incorporate the notion of reciprocal role relationships, as well as individual need-systems with which they are related.[30] With the development of such kinds of theory, emphasis upon phenotypical "types of leaders" and upon all-or-none differences between leaders and followers gives way to theories of genotypic role relationships among members, some or all of whom are viewed as performing differentiated leader functions. The nature of their role assignments varies both with individual properties (e.g., capacities and need-systems) and with group properties (group norms and objectives). It would be hard to say whether or not such a theory is more psychological than sociological. In any case it seems more adequately to account for the observed phenomena than any "merely psychological" theory.

The concept of role, as already indicated, is incomplete without the concept of *consensus*, which has to do with similarities among the individual expectancies which together make up the pattern of expectancies which constitutes a role. *Objective* consensus—as

[29] A notable exception, as early as 1936, is to be found in J. F. Brown, *Psychology and the Social Order*, New York: McGraw-Hill. Taking an avowedly Lewinian position, Brown's approach was that of group "membership character."

[30] Cf., for example, F. H. Sanford, "The Follower's Role in Leadership Phenomena" in *Readings in Social Psychology, supra*, and W. E. Henry, "The Business Executive: The Psychodynamics of a Social Role," *American Journal of Sociology*, 1949, 54, 286–291.

stressed, for example, by E. H. Sutherland in *The Professional Thief*
—may be described in terms of similarities of orientations (cognitive,
cathectic, and evaluative, as these terms are used by Parsons and
Shils) toward role-taking behaviors in a specified range of situations.
This, obviously, is a group property with which the behavior of
group members varies. An individual's role behavior may be expected
to vary more closely, however, with consensus *as perceived by him*.
This intervening variable at the individual level, perceived consensus,
is the analogue of objective consensus, at the group level, and the
relationships among the two are of considerable interest. The group
property of "pluralistic ignorance" (as F. H. Allport dubbed it in
1924), for example, can be described in terms of discrepancies between objective consensus and the distribution of members' perceived consensuses. Or, if individual behavior is taken as the dependent variable, discrepancy between perceived and objective consensus becomes a predictive variable of considerable power.[31]

Personality theorists, both psychiatric and psychological, have
long been interested in autistic distortions of objective consensus
(often referred to as "reality"). Two important contributors have
recently made such notions central to their theories: H. S. Sullivan
and N. Cameron. Sullivan, who insisted that the goal of therapy is
to bring into the realm of the "consensually validated" that which
had been "selectively inattended," apparently reached this position
almost without benefit of the sociological literature on "the social
self."[32] Cameron, however, is very explicit as to his sociological indebtedness, particularly to G. H. Mead. His concept of "public verification" seems the virtual twin of "consensual validation"; and his
demonstration of the psychotic's "pseudo-community" gives eloquent
testimony that psychological processes, in disturbed as in "normal"
personalities, are dependent upon a structured social world, however
illusory. Cameron's heavy reliance upon the role concept,[33] moreover, provides a particularly good example of the fructification of
psychological by sociological theory.

Objective consensus, in short, is a necessary cement for holding
social structures together. It is a composite group property, in the

---

[31] As illustrated, for example, in my monograph, *Personality and Social
Change*, New York: Dryden Press, 1943.

[32] Cf. L. S. Cottrell and N. N. Foote in *The Contributions of H. S. Sullivan*,
New York: Hermitage House, 1952, p. 190.

[33] See especially his *Psychology and the Behavior Disorders*, New York:
Houghton Mifflin, 1947.

sense that individual members contribute to it, and the individual contributions can therefore be studied at the psychological level. As a group property, moreover, it is related to other group properties and at the same time has effects upon individual properties—perceived consensus, among others. It is a measurable property *par excellence* for the study of the effects of social organization upon individual organization.

Such considerations of properties and structures, however, raise questions about processes. Descriptions of social and individual organization are, after all, cross-section views of states of affairs which, no matter how stable or how changing, represent continuing kinds of adaptation. At the social psychological and microsociological levels, it is molar behaviors of individuals which constitute the raw data for studying such processes of adaptation. Not molar behavior of separate individuals, however, but of interacting persons who are behaviorally interdependent; hence the relevant processes are conventionally known as those of interaction. But interaction as such is far too global a concept to be usefully applied to concrete situations. Interaction processes therefore need to be studied in ways which make it possible to distinguish among different determinants of behavior, and among different consequences, both for behavers and for observers of the behavior in question.

As of 1954, there seems to be no more useful approach to the study of interaction than that of *communicative acts,* defined as observable behaviors by which information, consisting of discriminative stimuli,[34] is transmitted from a human source to a human recipient. It is my present contention that an adequate theory of communicative behavior not only would necessarily draw upon theories both of social and of individual organization, but would also contribute to both. If so, it would provide an essential link in the chain of a unified science of man.

In the simplest possible communicative act one person (A) transmits information to another person (B) about something (X). (Though either the source or the recipient may be a collectivity, the "A to B re X" paradigm is still serviceable.) Under conditions of con-

[34] This definition is adapted from G. A. Miller, *Language and Communication,* McGraw-Hill, 1951: ". . . information is used to refer to the occurrence of one out of a set of alternative discriminative stimuli. A discriminative stimulus is a stimulus that is arbitrarily, symbolically, associated with some thing (or state, or event, or property) and that enables the stimulated organism to discriminate this thing from other things."

tinuing relationships between A and B, A and X, and B and X, it becomes necessary to consider all the possible relationships among them as together constituting a system. These relationships seem most readily handled in terms of orientations of cognitive, cathectic, and evaluative nature; they include not only A's and B's orientations toward each other, but also the orientations of each toward X and, further, the orientations of either person toward each of the other two elements as perceived by the other person. Such a communication system, as distinguished from an interaction system, includes the object of communication as well as the communicants.

The state of the system at any moment is an outcome of previous communicative acts, usually involving both the same system (i.e., the same A, B, and X) and other analogous systems in which two of the three elements are the same (e.g., A, B, and X; or A, C, and X). Communicative acts are instigated by the perception of the system as being less than optimal in view of the situational demands of the moment. More specifically, as a function of valences toward B and toward X, communicative acts are instigated by A's perception of incongruence between his own and B's orientations toward X, and are directed toward the achievement of congruence between his own and B's orientations toward X. As a function of the same variables, congruence of A's and B's orientations toward a given X varies with the frequency of their communication about X. The possible consequences of a given communicative act include changes in A's and B's orientations toward each other, as well as either's orientations toward X.[35]

Quite apart from the advantages and shortcomings of this particular model, it seems clearly possible both to conceptualize and empirically to observe group properties in ways which relate them to communicative behavior, which in turn is related to individual properties which are definable both conceptually and operationally. If, as appears to be the case, it can be shown that communicative behavior is determined in predictable ways both by group and by individual properties, and also that stability and change of both individual and group properties are determined by the nature of communicative acts—if so, then the path toward a unified science of man will be so much the smoother.

[35] The rationale of such a theoretical framework is more fully developed in my article, "An Approach to the Study of Communicative Acts," *Psychological Review,* 60, November, 1953, pp. 393–404.

Such a framework leans heavily upon the concept of consensus. It assumes, on the one hand, that stability of group life is impossible without some degree of common orientation toward the common environment (both human and other), and that the "social reality" which corresponds to such common orientations is created, maintained, and/or modified by communicative behavior. Social reality, including the rules for communicative behavior itself, is a group property (it is meaningless to regard it as a merely individual property), but it can exist only through the participation of individuals— both as "interiorizers" and as communicators. It is also assumed, on the other hand, that stability of individual human life is impossible without participation in a system of social reality, and that the individual learns to participate in it as he learns to communicate. This he does in a manner varying with his own idiosyncratic experience, so that consensus never represents identity of orientations and, moreover, there are individual variations in perceptions of the nature and degree of existing consensus. Individuals' communicative acts are thus determined by private versions of consensus, which at the same time tend to be "corrected" by the consequences of those same communicative acts.[36] In such ways as these are consensus and communicative behavior interdependent.

Such a framework is also consistent with the phenomena of roles, if a single additional assumption is made: that role behaviors may be viewed as communicative acts. (It is not necessary to assume that they are nothing but communicative acts.) The reciprocally supporting nature of interlocking roles seems, in fact, to require a theory of communication. Role behavior is necessarily oriented toward something, and is observed by others as such. Hence, whether by word or deed, the role-taker (if observed) is necessarily transmitting information to some one about something.[37] In particular, he is transmitting information concerning his own orientation toward that something, which is a necessary pre-condition if consensus is to be achieved. Roles, in short, represent social reality as applied to persons

[36] One of the interesting extensions of the psychological theory of reinforcement has to do with the conditions under which orientations of *both* behaver and observer are reinforced by the same act. Assumptions concerning consensus are necessary to account for this phenomenon.

[37] In order to fit role behavior into the A-B-X paradigm, above, it is sometimes necessary to consider either A or B as the object of communication—the X. Since it is possible conceptually to distinguish between persons as communicators and as objects of communication, this becomes only a special case of the general formulation.

viewed as position holders. Like other forms of social reality, they are maintained by consensus and communication; and role behaviors, like other behaviors which are influenced by social reality, owe their idiosyncratic forms in part to unique experiences of communication and unique perceptions of consensus.

V. Some Research Needs toward Better Articulation

The kind of articulation that will enable sociology and psychology to be maximally useful to each other will never come about by persuasion from either side, nor from "neutrals," nor even from those few individuals of whom it may be said "they have both skills in one skull." Neither will it come about by sheer blue-printing, no matter how rationally organized the proposed division of labor. A necessary condition for such articulation is empirical research which, having been planned with problems of both collective and individual organization of behavior in mind, presents convincing evidence that variables at one level have effects upon the other. The following suggestions, therefore, have been selected with an eye to possibilities of researchability, as well as to theoretical contributions toward articulation:

1. Much more extensive—and more imaginative—research is needed concerning social structure as a channeller of communication. This problem reappears in countless guises, and it is easy to construct plausible accounts of how existing structures maintain themselves by controlling who communicates to whom about what, and thus perpetuating consensuses which support the structure. There is, however, very little support for these plausible accounts which rests upon systematically obtained, empirical data. Laboratory studies[38] have tested several significant hypotheses concerning who communicates to whom about what with what effects, in the laboratory situation. Comparable studies in "natural" habitats are, of course, more difficult, but they are also more rewarding if one is interested in "natural" social structures. The goal of research in this area should be nothing less than an understanding of the conditions under which communication practices vary with social structure,

---

[38] Especially those by L. Festinger and his associates. Cf. *Theory and Experiment in Social Communication*, Research Center for Group Dynamics, University of Michigan, 1950. See also H. H. Kelley, "Communications in Experimentally Created Hierarchies," *Human Relations*, 1951, IV, 39–56.

change in either being studied as a consequence of change in the other. Such studies, particularly if so planned as to include data concerning individual factors with which the initiation, reception, and retention of communications are related, would be quite as useful psychologically as sociologically, as well as having articulatory value.

2. More adequate operational definitions of role are needed; our present poverty in this respect is paralleled by the paucity of systematic role research. Merely to describe, or even to measure the attitudes or behaviors of persons who are presumably taking a specified role is no substitute for testing hypotheses concerning specific dimensions of the role concept itself. Research programs in this area are beginning,[39] and there is reason to anticipate the development of a body of interrelated propositions concerning social structure, consensus of role expectations, role behavior, and personality structure. Such a development might well constitute the main line of theoretical integration between sociology and psychology.

3. If the concept of consensus has the importance which I have attributed to it, then it, too, needs both conceptual and operational refinement. Extremely little systematic information has been gathered as to differentiated aspects of consensus which are hypothetically related to either group or individual properties. At the very least, consensus regarding cathectic, cognitive, and evaluative orientations toward both human (individual as well as collective) and non-human "objects" must be investigated; and the conditions under which various kinds of consensus are closely or loosely interrelated need to be known. Such research presupposes the development of indices of consensus, non-verbal as well as verbal. Many of the hypotheses to be tested in such a program of research would include variables concerning role and concerning communication.

Among the several differences which have prevented a consolidated attack by sociologists and psychologists upon their central common problem—the behavior of individuals in a structured group setting—probably none has been more crucial than the psychologist's and the sociologist's proclivities for studying the same behavior

---

[39] Of particular interest are those by S. A. Stouffer, by T. R. Sarbin and at the Institute for Social Research, University of Michigan. In all of these programs, work has far outstripped published reports.

as functions of persons and of roles, respectively. Individuals' behaviors must be studied in each of these ways, and for certain purposes either way alone is all that is necessary. But for other purposes it is desirable that the same behaviors be amenable to study in either way. Toward this end both a common vocabulary and distinctive vocabularies are necessary—enough in common to make communication possible, and enough that is distinctive to serve as a reminder that what is to be learned from the social organization of behavior is not identical with what is to be learned from its individual organization. The present requirement for a unified science of man is not fusion of the different levels of inquiry, but recognition and understanding of what is identical and what is distinctive. From such recognition and understanding, theoretical integration may yet emerge.

CHAPTER 8

The Forward View

JOHN GILLIN

No ONE CAN SAY with certainty when or if we shall have a comprehensive and integrated science of social man. The authors of this volume hold the view that some movement in this direction, if soundly carried out, is desirable. No value is seen in tendencies that promote a continued splintering of the knowledge of mankind which we already possess and hope to increase. If one is primarily interested in the application of science to the solution of contemporary crises in human relations, an orderly mobilization of scientific resources would seem to promise greater results than scattered "firing at will." Although we view with reserve the image of science as the all-embracing and sole "savior" of man from his own errors, we regard sincere efforts to solve human problems by scientific means as legitimate and worthy.

On the other hand, if one is interested only in increasing *understanding* of men and their works, without attempting application to "practical" problems, it also seems indisputable that such an objective can be more fully and rewardingly achieved if knowledge of the field is comprehensively and coherently organized. A fuller, deeper, and more reliably based understanding of human life we also regard as a worthy goal in itself. All tribes and peoples have sought to understand man, the meaning of his existence, his place in the universe, his doings on this earth. Such seeking seems to represent a deep-seated and age-old yearning. Insofar as science, in addition to religion, aesthetics, philosophy, and other serious approaches to

257

these matters can help to provide answers, no effort should be spared in its behalf.

Whichever may be one's primary interest in a science of social man—"pure," "applied," or both—all will agree that theory of a certain degree of coherence and unity is of great value, if not indispensable, for fullest results. The probabilities are that the progress of all interests, whether of the "pure" or the "applied" types, can be measurably increased if carried on within a framework of common understandings.

What is theory? What useful purpose does it serve? What are the criteria of "good" or useful theory? To some readers these questions may seem elementary or naïve. Yet a perusal of the current theories in "social science" will show that frequently little attention has been paid by their authors either to general canons of scientific theory or to the scientific requirements of theoretical formulations in their own fields. The authors of this book are of the opinion that no mere splash of words, no matter how recondite, constitutes a viable scientific theory. But at the same time they are aware that many words may be required to elucidate a valid theoretical formulation.

To put the present position in its simplest form—every "normal" human being is, willy-nilly, a theorizer. This is merely to say that no normally intelligent member of the species Homo sapiens (with an I.Q. of, say, 90 or above) can look at anything without formulating a theory (or following some traditional formulation) about what he observes. To take a very trivial example, Joe Doakes hammers a nail into a board and the nail crumples. Doakes says to himself (if he has no audience), "That must be hard wood instead of soft wood." That statement of Doakes is a theoretical conclusion or theorem from certain implicit postulates. It is meant to explain the result he had from trying to drive the nail. Furthermore, it is a verifiable theorem. There are certain methods available to Doakes to determine whether or not the piece of wood involved is in fact northern oak or southern pine, or something else. Thus everyone theorizes.

This is supported by the records of ethnology, archeology, and history. There is no people known to us which has not had some body of mythology, theology, superstition, or elementary science designed to explain (and often to predict) the phenomena of nature and the fate of man. For example, we may deduce from the archeological evidence of burial trappings that even as early as the middle of the

Old Stone Age the hominids of that era held fairly definite theories of the fate of the individual after death.

A theory is, then, in one sense, a way of ordering one's thoughts about a given area of experience. Any theory offers one or a series of possible explanations of the matters under consideration, and on the basis of these explanations allows either implicit or explicit predictions about future events. "Theorizing" of some sort seems to be a universal characteristic of human beings and probably serves various psychological functions. *Some* sort of explanation is better than none at all or complete confusion. But, while organization of thought provides a degree of satisfaction for the moment, the long-run value of a theory resides in the reliability of its explanations and predictions, and generally speaking this means that it must be verifiable. Experience has shown that in point of reliability scientific theory has proved most rewarding for scientific purposes. Of course, for other purposes, such as religious and art appreciation, aesthetically and intuitively formulated theories may be very useful.

Perhaps in the present connection it is not superfluous to set down once again certain elementary features considered desirable in any scientific theory of social man, regardless of the formal structure which such a theory may take.

1. Terms or symbols used in stating the theory should be clearly distinguished between (a) primitive or undefined terms whose meaning is derived from their use in the context of the theoretical statements, but in no other way; and (b) definable terms. The latter should be unambiguously defined either verbally, operationally, by demonstration, or according to some other established method. One hardly needs to remind students of social science that misunderstandings over terminology have blocked many an attempt at theoretical integration.

2. The statements of "theoretical possibilities" may be set out in a variety of ways. In a formal logico-deductive-empirical system they usually appear as "postulates" or "assumptions." They may, however, legitimately be put forward as "propositions," "hypotheses," "theoretical insights," and so on. Whatever the form of their presentation, however, the statements of theoretical possibilities should be *logically related one to another,* and such relationship must be made as explicit as possible. Such a rule is, of course, an elementary precaution against confusion which might otherwise result from overlapping statements. It is also an essential guard against omission and

oversight. Only when one has logically integrated his statements of theoretical possibilities has he any reliable basis for judging whether or not he has overlooked certain other possibilities. Finally, the logical integration of theoretical statements of this kind is an essential prerequisite to the drawing of comprehensive deductions or logical consequences.

It is, of course, well known that theoretical progress in social science has been—and still is—blocked by neglect of this canon. Collections of loosely integrated or frankly "free floating" hypotheses and "hunches" have on occasion passed for theoretical "systems" in the behavioral sciences, usually to the infinite confusion of both their authors and their followers.

On the other hand, there is the type of "theorizer" who seems to think that logical integration is all that is required of a useful scientific theory. In fact, one can derive a high degree of aesthetic pleasure from arranging propositions in a logically correct and symmetrical form. This was one of the principal pastimes of the monastic scholars of the Middle Ages. But the modern scientific theorist always is aware that logic is only a tool for the ordering of theoretical possibilities that may ultimately be tested either directly or indirectly. Otherwise the making of a theory becomes nothing more than an intellectual exercise or a "juggling of concepts."

3. Among the aids to clear-cut, logically integrated theory is the principle of economy, which holds that each theoretical statement should be as "clean" as it is possible to make it; that is, it embodies or invokes no more than the minimum number of concepts required to state the highest degree of generalization appropriate to the theoretical possibility with which it is concerned. In other words, a theoretical system should be as "simple" as possible consistent with the analysis of the problems to which it is addressed. Systematic specification of concepts, or the symbols for them (as described in Point 1 above), is of great assistance in achieving "economy" in this sense.

4. If the "theoretical possibilities" of a system are arranged in any logical manner (and this has been advocated above), their logical consequences or derivations should be sought and made explicit. This again is an elementary precaution which helps one to guard against "overlooking anything." In a logico-deductive-empirical system such products are called "theorems," but even in less formal presentations they are not to be ignored. To make a long story short

in the language of the vernacular, every theoretical system should be "squeezed" to the utmost limit of its possibilities in order to make sure that all logical implications have been fully realized.

To be realistic, however, we must recognize that at a certain stage of many a scientific exploration or inquiry this condition cannot be met, for the reason that the field is not yet sufficiently familiar to the investigators to permit a comprehensive formulation of "possibilities." In such circumstances the student must often be content with *working hypotheses* or even vaguely formulated *hunches*. Such, often hazy, concepts are assuredly statements of "possibilities" and, if properly used, serve as *research leads* which aid in the further development of the scientific knowledge of the field in question. The verification or rejection (on empirical grounds) of a "working hypothesis" will usually yield data the implications of which can be theoretically formulated. By this process of alternately formulating working hypotheses, testing them against data, expanding the range of hypotheses, and so on, sufficient "knowledge" of a field of inquiry can usually be built up to the point where the various verified hypotheses may be connected together into at least the beginnings of a logical arrangement.

Thus, while the "working hypothesis" is an essential tool in the development of a science, the crucial necessity remains of connecting together such statements as soon as possible into a logically consistent body of theory. In social science particularly there has been a tendency for the individual investigator (especially in the United States, where individualism is highly valued) to be content with the "free-floating" hypothesis alone, or actually to capitalize upon it, making it in a sense his own private property without attempting the necessary effort to consolidate his findings in a general body of scientific theory.

5. From what has gone before it is of course evident that a scientific theory must be so framed that it is testable empirically—that is, by the evidence available either directly or indirectly to human senses. Thus a statement like, "I believe that the Spirit of Spring causes Love to blossom in the human heart," is not, without considerable elaboration, an empirically testable proposition. Among other deficiencies, "The Spirit of Spring," "Love," and "blossom in the human heart" have no stated empirical referents that are open to investigation. Of course, it is conceivable that such metaphors can be translated into scientifically empirical terms—but that is exactly what the formu-

lator of useful theory must do. He must forswear the use of fanciful figures of ambiguous meaning if they do not lend themselves to empirical verification. Of course, the scientific theorist has a perfect right to try his hand at poesy and all other forms of art so long as he does not lead his audience to believe that, while so doing, he is dealing with scientific theory.

In logico-deductive-empirical systems the postulates are usually tested through the theorems, which are logically derived from them. In the negative case, if the data show a theorem to be incorrect or only partially in correspondence with the evidence, the postulates (and perhaps the concepts and terms themselves) from which the theorem was deduced must be discarded or modified. On the other hand, empirical verification of a theorem will have the effect of adding validation to all other elements of a logically organized system. This is of course one of the virtues of logical organization, namely, that empirical tests have reverberations throughout the system, whereas in more loosely organized bodies of theory the remoter implications of test are not always so evident or so easily traceable.

Although a viable scientific theory should always clearly point to data that may be investigated, the theory itself, strictly speaking, is not required to specify or spell out the *method* and *technique* of investigation. Methods and techniques have frequently been devised only after the "problems" to be studied have been elucidated theoretically. Also it must be recognized that scientific discoveries of value have been made by sheer "play with technique," to be later elaborated theoretically; but experience has shown that, at least when a field of knowledge has been developed beyond the first exploratory stages, method and technique are as a rule more fruitful when theoretically guided than otherwise.

The usefulness of ordering or organizing our theoretical knowledge of social man systematically according to some such rules as those discussed has several aspects. First, a common framework enables A to see at once the significance of contributions and discoveries made by B . . . X. Second, a comprehensively organized frame of reference makes clear existing "gaps" in current knowledge and permits the planning of research with precision. Thus extension of the "frontiers" of knowledge can be pushed forward systematically, rather than in helter-skelter fashion and many theoretically unguided "false starts" can be avoided. Third, a rational organization of theory renders more efficient the mobilization of knowledge accumulated by various scientists and disciplines for the solving of new problems, whether of "pure" or "applied" science.

One caution should perhaps be mentioned. If what we have called "the science of social man" should ever achieve a systematically organized body of common theoretical understandings that represent consensus among the leading investigators in the field and claim their allegiance, such a system must never be allowed to become "authoritarian," in the sense that any deviation from the system is regarded as something akin to "heresy" which is condemned out of hand without examination of its merits. It is desirable to avoid intellectual chaos in the study of man, and a comprehensive framework of theoretical agreements can be a useful instrument to this end. But we cannot forget that much of scientific progress has been built on originality of insight and interpretation. If a scientist feels that he can demonstrate that his contribution is incompatible with the general body of theory, he must be given every opportunity to do so by scientific procedures. It is to be expected that the common body of understanding will be modified and revised from time to time by such demonstrations. At the same time true originality—as distinguished from mere verbal virtuosity and emphemeral bids for attention, for example—can be recognized, to the benefit of all.

In the preceding chapters we have attempted to show by historical review how certain concepts have arisen and how some of them have been discarded. And we have endeavored to indicate how, from cooperative work on certain common problems, a number of theoretical concepts and propositions have arisen that are now more or less the shared property of anthropology, psychology, and sociology, or, at least, are reasonably meaningful and useful in theoretical discussions across departmental boundaries. It would seem, from these reviews of the situation, that the possibilities for increased consolidation and systematic expansion of our knowledge of social man are promising.

As was stated earlier, it is not our aim in this book to offer even a tentative theoretical *system* covering those areas in which members of our three disciplines have found it helpful to work together. But it may be worthwhile to point briefly once again to some portions of the scientific landscape where gates are opening and the fences seem to be coming down. For present purposes, something like an informal "check-list" seems better suited than a formal analysis.

The Human Organism. There is at present general agreement among the sciences of social man that human beings are organisms and as such conform to certain scientifically formulated general principles governing all organisms. At the same time, the present species of man is a special kind of organism having certain characteristics common to the members of the species but not shared with other species. The concepts and propositions concerning the human organism have for the most part been formulated and validated by disciplines usually considered to comprise the group known as the "biological sciences." The theoretical problem for a science of social man consists in deciding which concepts and propositions from this body of science are necessary for an understanding of man's behavior considered on the social level, and how such concepts and propositions can be most usefully formulated (or "translated") for this purpose. As Newcomb points out, "If each of the several disciplines of human behavior took as its microcosm its neighbor's macrocosm, and its alternate neighbor's microcosm as its own macrocosm, their problems of articulation would be relatively minor ones." This method of articulation can, of course, be extended through the entire range of the "hierarchy of the sciences." Another way of putting this is to point out that it has often proved useful for a science at one level to incorporate among the *postulates* of *its* theoretical system certain statements from the theoretical system of a "lower" level which in the latter formulation have the status, not of postulates, but of verified *theorems* or "laws." Thus the statement that, "All living human beings require nourishment through the modification and incorporation into their own structures of organic and inorganic materials foreign to their own structures," may be regarded as a verified theorem or law of biology. In a formal science of social man this might be treated as a postulate. If this is true (and biology says that it is), then certain conclusions ("theorems") can be drawn regarding man's social behavior and its products. It should be clear that these procedures in theoretical formulation are quite different from "reductionism." In the case of a theory of social man, one form of "reductionism" would endeavor to explain man's social behavior *entirely* on the basis of certain propositions of the biological sciences that have been found to be true with respect to the individual organism, thus leaving out of account (and consequently leaving unexplained) many phenomena of the social level, which biology does not attempt to consider.

Some retardation in the development of our knowledge of social man has been caused by fundamental misunderstanding of this point of view. The literature contains many discussions that can only be considered "sham battles" waged against what has been inappropriately identified as reductionism. While it is generally agreed that true reductionism is a sterile approach to a science of human social life, it seems obvious that the latter must adapt to its *own purposes* the findings of other sciences that are significant for such objectives. It seems to be extremely poor strategy for those endeavoring to understand man in his social life to rule out any contributions that might be made to such understanding by biological, physical, or "natural" science. The problem for social science, as we have said, is one involving decisions regarding selection and significance.

Human Behavior. Obviously this does not refer to a simple concept or set of propositions. Nevertheless, a general theory of social man must involve some assumptions regarding the fundamental aspects of human behavior. Here again help must be had from the sciences of the organism—for example, with respect to the limits of behavior (what the organism can and cannot do), the presence or absence of significant hereditary factors, and so forth. Furthermore, psychology naturally contributes heavily. Some may say that a general theory of social man is impossible until psychology settles for itself certain problems of behavior. Are we to use Freudian theory, general learning ("behavioristic") theory, Gestalt or field theory, "perceptual learning" formulations, or what? Here, it seems, decisions again must be made concerning the types of questions regarding human behavior with which a general theory of social man must be prepared to deal. As a result of such decisions, it is entirely possible that concepts and propositions may be drawn from all of the current psychological theories. It may well be demonstrated that each of the present "schools" of psychology can contribute something to a general theory of social life. It is already apparent in some writing and research at the social level that various parts of the several psychological formulations can be used together. For instance, some workers have found the most useful guidance for response and motivational aspects of social behavior in terms of "learning theory," while at the same time "perceptual learning," for example, has offered more rewarding leads in those areas that have to do with stimulus. At all events, a general formulation will probably show that it is not necessary to accept or reject the contribution of certain schools *in toto,* but probably will demonstrate that there is something useful in each school's findings and formulations.

It is with such prospects in mind that we have ventured to assert that the development of a collaborative (or cross-disciplinary) science of social man does not carry with it the inevitable abolition of current disciplines, specialties, or "schools." Once a general understanding is established, it is quite possible that the various psychological schools can still maintain their integrity as systems of theory *for certain special purposes*. The general theory could well make use of selected contributions from the special theories without impugning the validity of either. Incorporation in a general theory will, of course, still leave any concept or proposition open to ultimate logical and empirical test. At the present time, as several of the preceding chapters have indicated in passing, there tends to be an "all or none" atmosphere in some circles in psychology: "value" theorists will have "nothing to do" with "drive reduction" psychologists; "field theory has nothing in common with learning theory"; and so on. Such statements, if taken literally, would call for an enormous wastage of scientific effort, that is, would require "throwing away" the entire results of theoretical and research work of the followers of the criticized school or coterie. Such rivalries can be reduced and the energies invested in them can be constructively channeled for the benefit of our knowledge of mankind as a whole. We believe that a collaborative science of social man will afford an instrument for achieving this goal. What has just been said about rival psychological theories applies with equal force to many "competing" schools in sociology and cultural anthropology.

Space is not available for a detailed discussion of the concepts which may be incorporated in a general theory in those phases dealing with "human behavior." The following words would probably evoke a fair degree of common understanding, if not of complete agreement: learned and unlearned behavior, stimulus, perception, stimulus situation or configuration, response, reaction, drive, motivation, instigation, reward, reinforcement, action, internalized behavior, cathexis, projection, identification, displacement, life space or behavior space, symbolic behavior, goals, valences, values, generalization, discrimination, extinction of habit, and so on. No complete list is attempted here.

As both Smith and Newcomb have pointed out, certain psychologists have become increasingly aware of the fact that the individual's behavior cannot be realistically considered only in isolation and have accordingly given attention to the interactional and social aspects in their theoretical formulations.

Interaction. As the study of man has progressed, it has become evident that neither a knowledge of the organism alone nor of the psychologically isolated individual is sufficient for an understanding of the ways in which human beings actually live and behave. All normal human beings react with others of their own kind. Much of the validated theory regarding interaction has been developed by sociologists, although as the authors have pointed out on previous pages, not always in line with accepted psychological principles of human behavior. Sociologists and others, however, have demonstrated certain general "types" of interaction which seemingly recur throughout the species: for example, dominance-submission, leadership-followership, instigative-passive, originating-terminating, balanced-unbalanced, and so forth. This incomplete list of words does not represent mutually exclusive categories.

Grouping. The theory of groups has likewise been largely a product of sociological research and theory. It is a demonstrable fact that human beings live, behave, and interact in groups, a considerable variety of which have been identified and classified throughout the human species. Everywhere we find groups organized on the basis of sex, locality or territory, age, and "special interests"—such as congeniality, occupation, religion, economics, politics, aesthetics, and so on. Such a classification of groups implies group *goals*. Both the theory of goals and the theory of *group structure* are in need of further elaboration and test.

Culture. As Becker shows, the notion of culture has undergone a long series of vicissitudes. Yet the concept and its theoretical elaborations are mostly the product of cultural anthropologists. The evidence shows that although man is a certain type of organism with a wide range of capacities within determined limits, although he behaves in consonance with certain "psychological" laws or principles, although he is interactional in his most significant behavior, and although such interactional behavior takes place in groups, we must still consider the fact that the behaviors exhibited in any case follow certain "patterns" common to the members of specifiable groups or categories, but not necessarily universal to the species. The latter have been labeled "culture patterns."

As Smith points out, "culture" is a concept rather than a theory;

and all would agree that considerable elaboration of cultural theory is required before we shall have anything approaching a comprehensive science of social man. Yet the culture concept and many of its corollaries have become the common property of various sorts of social scientists. Most sociologists and psychologists would feel that they have some understanding of such words as culture trait, complex, culture area, ideal pattern, behavior pattern, acculturation, material "culture," cultural integration, and so on. As in the other disciplines, controversies which have occupied anthropological theorists have involved "realists" vs. "nominalists," "historicists" vs. "functionalists," "idiographers" vs. "nomotheticists," and so on. It is the contention of this writer that in most cases these supposedly opposing points of view or approaches do not represent "either-or" choices for science. It may be necessary, however, to restate some of the theoretical propositions in mutually understood terms in order to bring to light certain internal consistencies which actually exist in what on the surface appear to be antithetical approaches. A general framework of theory for the study of social man would appear to be a setting in which this could be accomplished. Murdock cogently argues that adequate cultural theory cannot be based on data derived only from Western civilization and shows that the anthropologists' accumulated knowledge of hundreds of other cultures must be utilized in the enterprise.

Certain published discussions would make it appear that the historical study of cultures is in some ways incompatible with a generalizing study of culture. Yet there is no *a priori* reason or conceptualized empirical evidence to indicate that generalizations tested on contemporary cultures should not likewise apply to those of history. Nor is there reason in the contention of some "generalizers" that, because certain cultural propositions cannot be directly tested against the data of history, all historical study of culture is worthless. It is, of course, true that, as in other disciplines, some anthropologists are temperamentally drawn to one type of approach, while others find some other point of view more congenial to their personality structures. But we must be continually on guard against confusing temperamental preferences of working scientists with the worth or validity of their theories.

Since "culture" in one accepted sense covers the whole range of human social life, cultural studies at the level of "culture content," at least, must inevitably come into contact with the work of other specialists. For example, no cultural theory that deals with "content" can rationally ignore the findings and formulations of students of technology, economics, religion, politics, art, and other special aspects of cultural manifestations. Likewise, that part of cultural theory that deals with cultural behavior (whether manifest or implicit) cannot usefully remain oblivious of the work of psychologists. On the other hand, those studies of culture that are carried on at a high level of abstraction, dealing with "pure patterning" in its many theoretical aspects may, in the division of labor, be the exclusive work of cultural theorists. There seems to be no rational reason, however, why even such theoretical work should not be so stated as to be accessible to other students working on somewhat lower levels or with "real human beings." In fact, if we follow the principles of useful theory sketched out above, any formulations on the purely abstract level which are intended to appeal to data rather than to plausibility, must, through their derivations and consequences, lead to statements that can be empirically tested in some way or other.

Thus cultural theory would seem to lead ultimately in many directions to contacts with various specialities.

Social Structure. As Parsons, among others, points out, the organization and composition of societies and groups can for certain purposes be considered apart from their culture. There have been decided tendencies, however, for theories of social structure and theories of culture to "meld" or articulate with each other, a trend that may lead to accepted consolidation of theory regarding these two aspects of social man. Likewise students of personality and of social psychology have found it increasingly difficult to deal with their materials without reference to social structure. "Reference group" theory represents one among several statements of problems which, starting from formulations of social structure, cuts across boundaries into both psychology and anthropology.

Personality. As Hallowell's chapter reveals, "almost everyone" is getting interested in personality. Each of our review chapters treats of this focus of interest in some respect. Once considered merely as an aspect of individual psychology, the study of personality now also claims the attention of anthropologists and sociologists in collaboration with psychologists and psychiatrists. In fact, this field is one of those in which members of the three disciplines have found it easiest to achieve a common focus and to "come to terms" with each other. Hallowell's review indicates that we shall probably see in the future an increasingly systematic and rigorous approach added to the clinical method which has hitherto prevailed. Any general theory of social man will inevitably treat of personality and the interplay of social and cultural influences upon it.

Symbolization and Communication. Practically all students of man would concur in the notion that an essential feature of his social life is intercommunication by means of symbols. Although linguistics has studied those systems of verbal symbols which we call language on a cross-cultural basis, and semantics has studied "meaning" largely within the context of our own culture, they have had little in common theoretically. Furthermore, work is still in its initial stages on a *general* theory of symbolization. It is desirable to know more about how the process works in the organism and in the individual. And students are becoming increasingly aware that not only language and explicit symbols but practically all aspects of human social life have meanings, often of great subtlety, whose definitions vary with culture and social structure. Certain concepts such as "expectancies," "cultural themes," "definition of the situation," "orientations," and "cultural premises," for example, will probably increase in scope and clarity when used in the context of a general theory of meaning. Such a theory would also articulate with certain psychological concepts and propositions relating to perception, attitudes and other "internalized" responses, secondary reward, and so forth; in fact, all that psychology has to offer concerning social behavior must be carefully examined and tested in this regard. And, of course, sociology and cultural studies must be relied upon to state the social conditions and patterning under which symbolization takes place or is performed.

Problems of communication are inextricably entwined with those of symbolization and meaning in a theory of social man, because symbols and meanings that remain private to the individual are in themselves alone of no significance to social life (although, as in the case of individual deviants, they may become so in the light of a general theory of communication). On the basis of present knowledge, a *sine qua non* of human social life is the communication of meanings from one individual to another by means of symbols. The means and channels of communication as they relate to social structure and to cultural systems considered as wholes (in addition to what are ordinarily described as specialized patterns and artifacts of communication in a given culture) have not been thoroughly explored either empirically or theoretically. It seems unlikely that any one of the disciplines as we now know them will be able to do the job. One of the urgent "practical" reasons for getting on with the task without delay in the present era lies in the hope that thereby we may eliminate or alleviate many of the intra-societal and international *misunderstandings* which lead to conflict and unrest.

Even so brief and incomplete a discussion of scientific problems respecting social man as we have offered in this book seems to indicate that strict isolation between disciplines studying man simply cannot be maintained, if certain questions are to be answered. Apparently the students of man and his ways must either stop being interested in such questions, or they must get together to some extent in order to help each other to provide reliable answers. In fact, as the preceding chapters show, this has actually occurred in numerous cases, and the results have usually been not only mere interstimulation of the scientists themselves, but an increase in knowledge that could not have occurred if disciplinary boundaries had not been breached. In fact, in certain "advanced" circles, "interdisciplinary" has by now become an old-fashioned word; it is no longer argued about—it is taken for granted. Nevertheless, this is not yet true of the great majority of the students of social man. And, in any case, there is still need for systematic consolidation of general theory covering the fields where specialists meet and for improved facilities of communication that would enable them to talk to each other with greater ease and understanding. What can be done about it?

One procedure is to let "nature take its course." In the long run, the mere fact that so many problems require the intermeshed use of distinct approaches will probably force the scientists involved to work out common theories for their own convenience. Team research, as it is sometimes called, does require some common understanding among members of the team, and for this reason has been offered by some proponents as the ideal or only effective incubator or "forcing bed" of integrated theory. Team research does not, however, inevitably contribute to the consolidation of general theory. In some cases, the common understandings reached between the team members concern only the problem at hand without regard to their more general implications. The urgency, financial and other, to finish a given job of research within a time limit and to produce "results" also frequently requires a team to agree upon an *ad hoc* formulation of its problem and to interpret the research data in terms of it.

Thus researches by a number of teams may produce a variety of theoretical formulations which, although they may be "interdisciplinary," are not necessarily generalizable, and which consequently fall short of their full usefulness. There is no need to cite any of the numerous examples of such results here. With the passage of time and through trial and error such matters may, of course, be "ironed out" without systematic attempts toward integration, but probably at the cost of considerable effort that could be avoided and at a price of delayed development of useful knowledge. Joint research on a common problem or problems through the team method involving different types of specialists will undoubtedly evoke awareness of common interests between the specialists involved and is, in fact, the only effective way of attacking certain kinds of problems. On the whole it is certainly superior to uncoordinated attacks by theoretically isolated and insulated individuals. Yet the problem still remains of insuring common understandings between teams.

Several other types of "bridge-building" devices intended to facilitate the approach of different kinds of scientists to each other have been briefly discussed in the first chapter—cross-disciplinary institutes, faculty seminars, conferences, and the like. Here we may mention one or two possibilities that have not as yet been tried out in the field of social studies.

Since difficulties involving terminology and its proliferation seem to absorb much energy and some emotion that could be more effectively used in the actual advance of knowledge, some means of systematically handling theoretical terms might be devised which could be available to all interested parties. One possibility would be a Bulletin of Scientific Terminology and Concepts in the Social Studies. To such an organ all authors and investigators in the field of social man would contribute such words or other conceptual symbols that they are originating or using in original or previously unestablished ways. Thus, concurrent with the publication of an article, monograph, or book employing any new or unusual terminology the author would contribute a list of such items to the Bulletin. The verbally defined terms would be accompanied by short verbal definitions, the operationally defined items by brief descriptions of the operations specified, undefined terms would be listed with reference to the published work where they appear, terms defined only in the system of the author would likewise be accompanied by reference to his work, and so on. A part of the Bulletin could be given over to articles discussing terminology, comparing usages, offering cross references to various authors, and the like. Such a Bulletin could be produced in such a way that the subscribers could cut out and file the items for their own use and ready reference. For example, binding at the top of the page and printing on only one side might be one mechanical solution.

Some such clearing house would serve the purpose of keeping the scientific public in the field *au courant* concerning new words and usages and would clearly expose the latter for systematic comparison with others. Discussions in the journal would, of course, inevitably deal with concepts as well as mere verbal usages. Any worker in the area of social man would thus be relieved of the necessity of searching vast amounts of literature, particularly in disciplines other than his own, when preparing the terminology to be used with his own research and theoretical material. Furthermore, authors of new contributions would have a medium in which terminological matters could be discussed without necessarily obscuring the merits of their theoretical offerings *per se*. As things stand now, a review or criticism of a new contribution frequently takes as its exclusive target the author's terminology with the result that sight is lost of the new ideas, if any, he has to offer. Such a journal or bulletin, in short, would seem to contribute to mutual understanding both within and across disciplines, and might lead to some further standardization of terminology or, at least, further consensus regarding usages, which in turn would facilitate the wider comprehension of concepts.

A standing, but working, joint committee on terminology and concepts in the science of social man, composed of theoretically oriented members of the various disciplines would also be an aid in this matter. It must be understood that we have in mind the promotion of mutual *understandings,* not necessarily *uniformities.* Our basic notion in this respect is that specialists of different kinds need to devote all their energies to increasing our understanding of mankind, rather than diverting a goodly part of such energies to arguing at cross purposes over mutual misapprehensions that can and should be cleared up. If we can once "fight our way out" of the tangle of terminological undergrowth, we may be able to see the trees—and even the woods. Whatever devices that may contribute to such unobstructed vision we feel should be carefully explored and tried out.

The use of mathematical models, symbolic logic, and/or other types of "neutral" symbolization is advocated in some quarters as means to increased precision of theoretical discussion in a science of social man, at least in relation to certain kinds of problems. A systematic, concerted exploration of all such possibilities would seem to be indicated.

In addition to intellectual and technical difficulties which block the way to the effective mobilization of the necessary scientific resources for a fuller understanding of the problems of social man, there are probably certain factors of a psycho-socio-cultural nature. There has been no systematic field study of the internal "culture and social relations" of social science and of the motivations developed through their influence in the individuals involved therein.[1] As Murdock has asserted on an earlier page, science is a part of our culture. A careful and unemotional self-analysis of social science by social scientists themselves might prove to be a useful project for interdisciplinary team research. Until such thorough documentation can be produced, we have to rely upon impressions in which many observers appear to concur.

In the first place, scientists like other men differ in temperament and personality structure. There are and probably always will be individuals to whom interdisciplinary collaboration is uncongenial —and the authors of this book do not propose to dragoon such scholars into forced labor on a science of social man. Their work can still be of great value while carried on within the confines of their own disciplines.

The structure of academic organization doubtless has some inhibitory effects upon collaborative work. Most universities are divided into "departments" which in turn develop vested interests in salaries, personnel, budgets, and programs that the departmental members are naturally loath to relinquish or see "submerged." In these circumstances a young scientist wishing to "get ahead" is usually surest of recognition and promotion if he stays within his own department and his own profession. Apart from these considerations, the maintenance of departmental entities is justified by the

---

[1] Suggestive analyses of the social roles of academicians from the point of view of the sociology of knowledge are: Florian Znaniecki, *The Social Role of the Man of Knowledge,* New York: Columbia University Press, 1940; Logan Wilson, *The Academic Man,* New York: Oxford University Press, 1942. A short analysis of some of the professional socio-cultural aspects of social science in Latin America is in John Gillin, "Las Ciencias Sociales en America Latina," *Ciencias Sociales,* Washington: Pan American Union, 1953.

fact that certain types of specialized study can most effectively be carried on in a situation organized for such purposes. But many individuals with skills and training that would fit them for collaboration with colleagues from other departments either fail to develop the motivation to do so or are discouraged by the frustrations and low rewards. Here the essence of the problem seems to lie in administrative skill and reorganization that provides channels and rewards for selected scientists whose interests and capabilities fit them for interdisciplinary work, without at the same time destroying the integrity of the traditional departments, insofar as the latter are scientifically and educationally justified. As mentioned in the first chapter, this is being done in a variety of ways, although on a comparatively small scale. Experience in such cases seems to make it clear that the interests of both the individual discipline and cross-disciplinary field can be adequately safeguarded without the necessity of creating emotional and semi-political "issues."

One aspect of traditional training in anthropology, psychology, and sociology might be improved if we are to see more constructive and comprehensive work on a theory of social man. I refer to the fact that at the graduate level the student often receives more instruction in how to criticize polemically the theories of others than in how to construct theories himself and submit them to crucial test. A disproportionate amount of time is spent in the critical review of "the history of theories" in comparison to time devoted to learning how an elaborated and testable theory can actually be put together. Thus many a young doctor in the social sciences can tell you at length "what's wrong" with all the writers in his field from Plato to the present day, but his notions about how he himself could formulate more useful theory are often confined to some rather hazy notions about unconnected hypotheses and how to phrase plausible "arguments" in polemical style.

A thorough, critical knowledge of previous work is, of course, a prerequisite to systematic scientific advance, but is not to be regarded as an end in itself. In this connection, it would seem that all professional candidates in the sciences of social man should be required to have some training in logic and "the theory of theories."

Finally, we must remember that in Western society scientists like others live by a culture that highly values "individual distinction" and offers "competition" as the most generally available pathway toward it. Since a social scientist can seldom by the practice of his profession win "distinction" measured in monetary terms, it is not surprising that other ways are sought by some. Often the desired goal seems to be defined as "being different from others," that is, being "distinct" rather than "distinguished" in other senses. So far as work on consolidated theory of human behavior is concerned, this tendency on occasion has had the effect of cluttering the field with a variety of ostensibly theoretical statements that are purposely (if unconsciously) irreconcilable. The authors of such theories, together with their respective students and friends, can make careers of "defending their positions" and "counterattacking the opposition" without ever adding appreciably to our shared scientific knowledge. I know of no such writers who are consciously mountebanks, although some may be what Aldous Huxley has described as "adrenalin addicts" (never happy except in a fight). In most cases, it is alleged, the fault is to be found mainly with the competitive urge for distinction as structured in our "culture of social science." Insofar as such patterns and behaviors exist, it is obvious that professional organizations and scientific bodies can improve the situation partly by redefinition of "distinction." It must perhaps be remembered that we live in a culture which also values "cooperation with others" and "self-discipline."

Thus we must admit that there are difficulties, technical and of a psycho-socio-cultural nature, along the way toward the achievement of general theory and wider collaboration among specialists in the study of social man. But we do not regard such difficulties as insuperable. Already we have a vast amount of data and understanding concerning mankind. On the basis of the science we already have we know that time after time men have reached their goals in the face of what seemed to be insurmountable obstacles. It would be ironic indeed if we as scientific students of man should in our own case remain paralysed in the face of difficulties that are actually rather trivial and that we should thereby be blocked from our goal—which, after all, is nothing more and nothing less than a reliable understanding of social man, an understanding constantly increasing in breadth, in truth, and in clarity.

THE AUTHORS

Howard Becker is Professor of Sociology in the University of Wisconsin and former chairman of the Department of Sociology and Anthropology there. He is past president of the Mid-West Sociological Society. Among his books are *Through Values to Social Interpretation, German Youth: Bond or Free, Social Thought from Lore to Science* (with H. E. Barnes), and *Systematic Sociology* (with Leopold von Wiese).

John Gillin is Professor of Anthropology and Research in the University of North Carolina. He has written *The Ways of Men, Cultural Sociology* (with J. L. Gillin), and numerous monographs and articles dealing with anthropological field work and theory.

A. Irving Hallowell is Professor of Anthropology in the University of Pennsylvania. He is a past president of the American Anthropological Association, the American Folklore Society, and the Society for Projective Techniques, and was formerly chairman of the Division of Anthropology and Psychology of the National Research Council. Since 1951 he has been editor of the Viking Fund Publications in Anthropology. He is the author of *The Role of Conjuring in Saulteaux Society* and many other contributions to the literature.

George Peter Murdock is Professor of Anthropology in Yale University and former chairman of the Department of Anthropology at Yale. He has served as president of the American Ethnological Society and the Society for Applied Anthropology, and is past vice-president of the American Anthropological Association. He is the author of *Our Primitive Contemporaries, Social Structure*, and other books and monographs. Dr. Murdock was the originator and founder of the well known Cross Cultural Survey (now expanded into The Human Relations Area Files, Inc.).

Theodore M. Newcomb is Professor of Sociology and Psychology in the University of Michigan, in charge of the inter-departmental program in social psychology. He is a former president of the Society for the Psychological Study of Social Issues and of the Division of Personality and Social Psychology of the American Psychological Association, as well as council member of the American Sociological Society. His books include *Social Psychology, Personality and Social Change*, and *Experimental Social Psychology* (with G. and L. B. Murphy).

Talcott Parsons is Professor of Sociology and chairman of the Department of Social Relations in Harvard University. He is a former president of the American Sociological Society and of the Eastern Sociological Society. He is the author of *The Structure of Social Action, Essays in Sociological Theory, Pure and Applied, Toward a General Theory of Action* (with others), *The Social System,* and *Working Papers in the Theory of Action.*

M. Brewster Smith is a member of the staff of the Social Science Research Council, New York City, and was formerly Professor and chairman of the Department of Psychology in Vassar College. Since 1951 he has been general editor of the *Journal of Social Issues.* He is joint author (with S. A. Stouffer, et al.) of *The American Soldier,* Vol. 2, *Combat and Its Aftermath,* and of a number of articles on psychological theory and research.

READINGS

Allport, Gordon W., *Personality: A Psychological Interpretation*, New York: Holt, 1937.
A classic presentation of personality theory from a non-psychoanalytic standpoint, emphasizing the uniqueness of personality.

Asch, Solomon E., *Social Psychology*, New York: Prentice-Hall, 1952.
A textbook that subjects current assumptions to critical examination from the point of view of Gestalt psychology. Contains a critical discussion of cultural relativism.

Barnett, H. G., *Innovation: The Basis of Cultural Change*, New York: Mc-Graw-Hill, 1953.
Though written by an anthropologist, this original and ambitious work makes central use of psychological theory and research, particularly in regard to perceptual and thought processes. Its sophisticated use of psychological concepts is a model of effective borrowing from an adjacent field.

Benedict, Ruth, *Patterns of Culture*, Boston and New York: Houghton Mifflin, 1934.
A classic configurational interpretation of cultural "ethos," with psychological overtones.

Bidney, David, *Theoretical Anthropology*, New York: Columbia University Press, 1953.

Blum, Gerald S., *Psychoanalytic Theories of Personality*, New York: Mc-Graw-Hill, 1953.
A summary of orthodox and variant psychoanalytic theories of personality development, with notes referring to objective studies that bear on them.

Cooley, C. H., *Human Nature and the Social Order*, New York: Charles Scribner's Sons, 1902.

DuBois, Cora., *The People of Alor*, Minneapolis: University of Minnesota Press, 1944.
A monographic study of personality and culture in Indonesia, with independent expert interpretation of life histories and projective test records.

Durkheim, Emile, *Division of Labor in Society*, Glencoe, Ill.: The Free Press, 1947.

Durkheim, Emile, *Elementary Forms of the Religious Life*, New York: Macmillan, 1915.

Durkheim, Emile, *Suicide: A Study in Sociology,* Glencoe, Ill.: The Free Press, 1951.

Freud, Sigmund, *Civilization and Its Discontents,* London: Hogarth, 1930.
A discussion of the control of sex and aggression and the formation of positive social ties through internalization and identification. Carries to its logical conclusion the social implications of a motivational theory based on the reduction of tension.

Freud, Sigmund, *Ego and the Id,* London: Hogarth, 1927.

Freud, Sigmund, *New Introductory Lectures on Psychoanalysis,* New York: Norton, 1923.

Freud, Sigmund, *Problem of Anxiety,* New York: Appleton-Century, 1943.

Freud, Sigmund, *Totem and Taboo.* In *The Basic Writings of Sigmund Freud,* New York: Modern Library, 1938.
Freud's original attempt to treat primitive culture in psychoanalytic terms.

Fromm, Erich, *Escape from Freedom,* New York: Farrar and Rinehart, 1941. A neo-Freudian interpretation of personality and culture, applied to the case of Nazi Germany. Develops the concept of social character.

Gillin, John, *The Ways of Men,* New York: Appleton-Century-Crofts, 1948.

Haring, Douglas G., ed., *Personal Character and Cultural Milieu,* rev. ed., Syracuse: Syracuse University Press, 1949.
A collection of readings and bibliography.

Hilgard, Ernest R., *Theories of Learning,* New York: Appleton-Century-Crofts, 1948.
Expository and critical discussion of the principal contemporary psychological theories of learning.

Hull, C. L., *Principles of Behavior: An Introduction to Behavior Theory,* New York: Appleton-Century, 1943.

Kardiner, Abram, *The Individual and His Society,* New York: Columbia University Press, 1939.
A pioneering neo-Freudian attempt to interpret personality and culture relations in two primitive societies, in collaboration with the anthropologist Ralph Linton. Introduces the concept of basic personality structure.

Kardiner, Abram, et al., *Psychological Frontiers of Society,* New York: Columbia University Press, 1945.
A sequel to *The Individual and His Society* presenting neo-Freudian psychocultural analyses of three additional societies, including Alor and our own.

Klineberg, Otto, *Social Psychology,* New York: Holt, 1940.
A textbook that draws heavily on ethnographic materials. Contains an excellent survey of the cross-cultural constants of human motivation.

Klineberg, Otto, *Tensions Affecting International Understanding,* New York: Social Science Research Council, Bulletin 62, 1950
Contains a critical review of the literature on national character, with particular attention to methodological problems.

Kluckhohn, Clyde, Murray, Henry A., and Schneider, David M., eds.,

Personality in Nature, Society, and Culture, rev. ed., New York: Knopf, 1952.
A collection of readings, with introductory chapters presenting a theoretical framework.

Koffka, K., *Principles of Gestalt Psychology,* New York: Harcourt, Brace, 1953.

Kroeber, A. L., *Anthropology,* New York: Harcourt, Brace, 1948.

Kroeber, A. L., ed., *Anthropology Today,* Chicago: University of Chicago Press, 1953.
An "encyclopedic inventory" of anthropology prepared in connection with the international symposium sponsored by the Wenner-Gren Foundation. Chapters on "culture, personality, and society," "national character," and "the relation of language to culture" are particularly relevant.

Kroeber, A. L. and Kluckhohn, Clyde, *Culture: A Critical Review of Concepts and Definitions,* Cambridge: Papers of the Peabody Museum, Vol. XLVII, No. 1, 1952.
An exhaustive survey and analysis of definitions of culture and interpretations of the concept of culture.

Kroeber, A. L., *The Nature of Culture,* Chicago: The University of Chicago Press, 1952.

Lewin, Kurt, *A Dynamic Theory of Personality,* New York: McGraw-Hill, 1935.

Lewin, Kurt, *Field Theory in Social Science,* D. Cartwright, ed., New York: Harper, 1951.
A collection of theoretical papers by the late exponent of field theory, in which his approach is applied to social psychological problems.

Lewin, Kurt, *Principles of Topology Psychology,* New York: McGraw-Hill, 1936.

Linton, Ralph, *The Cultural Background of Personality,* New York: Appleton-Century-Crofts, 1945.
Introductory lectures in which, among other things, the concept of status personality is introduced.

Linton, Ralph, *The Study of Man,* New York: Appleton-Century, 1936.
An influential textbook of anthropology that introduced the concepts of role and status.

Lowie, Robert H., *The History of Ethnological Theory,* New York: Farrar and Rinehart, 1937.

Malinowski, Bronislaw, "Man's Culture and Man's Behavior," *Sigma Xi Quarterly,* XXIX, 170–196; XXX, 66–78. 1941–1942.

McDougall, William, *Introduction to Social Psychology,* rev. ed., Boston: John W. Luce, 1926.

McDougall, William, *The Group Mind,* New York: G. P. Putnam & Son, 1920.

Mead, S. H., *Mind, Self and Society,* Chicago: University of Chicago Press, 1934.

Mead, Margaret, *From the South Seas. Studies of Adolescence and Sex in Primitive Societies,* New York: Morrow, 1939.

Collects in one volume her well known books on Samoa, Manus, and the Arapesh, Tchambuli, and Mundugumor.

Merton, Robert K., *Social Theory and Social Structure,* Glencoe, Ill.: The Free Press, 1949.

Miller, N. E. and Dollard, John, *Social Learning and Imitation,* New Haven: Yale University Press, 1941.

Presents a simplified version of the behavioristic learning theory of Clark Hull, and applies it to experiments on imitation and to the anthropological concept of diffusion.

Murdock, George P., *Social Structure,* New York: Macmillan, 1949.

Newcomb, Theodore M., *Social Psychology,* New York: Dryden, 1950.

Parsons, Talcott and Shils, Edward A., eds., *Toward a General Theory of Action,* Cambridge: Harvard University Press, 1951.

Parsons, Talcott, *Structure of Social Action,* Glencoe, Ill.; The Free Press, 1949.

Parsons, Talcott, *The Social System,* Glencoe, Ill.: The Free Press, 1951.

Parsons, Talcott, *Working Papers in the Theory of Action,* Glencoe, Ill.: The Free Press, 1952.

Sapir, Edward, *Selected Writings of Edward Sapir in Language, Culture, and Personality,* D. G. Mandelbaum, ed., Berkeley: University of California Press, 1949.

A collection of classic papers including some of great influence in giving direction to research on culture and personality.

Sargent, S. Stansfeld and Smith, Marian, eds., *Culture and Personality.* Proceedings of an Interdisciplinary Conference held under the auspices of the Viking Fund, Nov. 7 and 8, 1947, New York: Viking Fund, 1949.

Papers and discussion by anthropologists, psychologists, and psychiatrists.

Spencer, Herbert, *Descriptive Sociology,* 16 vols., London: Williams and Norgate, 1873–1934.

Spencer, Herbert, *The Principles of Sociology,* 3 vols., New York: Appleton, 1896–1897.

Sullivan, Harry S., *Conceptions of Modern Psychiatry,* 2nd ed., Washington: The William Alanson White Psychiatric Foundation, 1947.

An influential psychiatric theory, related to psychoanalysis, emphasizing interpersonal relations.

Sumner, William G., *Folkways,* Boston: Ginn, 1906.

Sumner, William G. and Keller, Albert G., *The Science of Society,* 4 vols., New Haven: Yale University Press, 1927.

Thompson, Clara, *Psychoanalysis: Evolution and Development,* New York: Hermitage House, 1950.

An excellent introduction to psychoanalytic theory by way of an historical account of its vicissitudes from its inception to the present. The author is a neo-Freudian affiliated with Harry Stack Sullivan's views, but achieves a perspective relatively free from dogmatism or sectarian controversy.

Tolman, E. C., *Purposive Behavior in Animals and Men,* New York: Appleton-Century, 1932.

Weber, Max, *Theory of Social and Economic Organization,* translated by Talcott Parsons, New York: Oxford University Press, 1947.

Werner, Heinz, *Comparative Psychology of Mental Development,* rev. ed., Chicago: Follett Publishing Co., 1948.

A work in the European tradition emphasizing levels of organization and development. Children and "primitives" are compared.

Whiting, John W. M. and Child, Irvin, *Child Training and Personality,* New Haven: Yale University Press, 1953.

A pioneering cross-cultural study testing hypotheses from psychoanalysis and learning theory through the systematic rating and classification of existing ethnographic reports.

Woodworth, Robert S., *Contemporary Schools of Psychology,* rev. ed., New York: Ronald, 1948.

A simple and judicious introduction to the principal movements within twentieth-century psychology.

INDEX OF NAMES